Brain Mind Self

Brain Mind Self

Framework to Sustain a Healthy Self

Adrianne B. Casadaban, Ph.D.

Performance And Realization
Lafayette, CA

BRAIN MIND SELF:
Framework to Sustain a Healthy Self

ISBN: 1979928975
ISBN-13: 978-1979928977
Library of Congress Catalog Number: Pending

Published in the United States of America by
Performance And Realization
PerformanceAndRealization.com
Lafayette, CA 94549

Available from Amazon.com, CreateSpace.com, & Kindle

Contact Author: (925) 946-9991
BMSAI@protonmail.com
PerformanceAndRealization.com
BrainMindSelf.com

To Craig, Kimberley, and the rest of my family.
You are wonderful.
Words cannot express my gratitude.
We did it together!

Kimberley Dorothy Casadaban Schroder significantly contributed to the crafting of this book through her talents and dedication.

Table of Contents

Preface

"If we truly are discovering the way it is in nature, then we are all encountering the same phenomena." Words to that effect were Dr. Alan Schore's response to me when I first discussed my ideas with him years ago. Go ahead and publish it, he said.

This journey actually started for me what seems like eons ago, when Sister Joan Roach opened a whole new world to a sheltered Louisiana high school girl. She said we could do our physics class science project on anything we wanted, so I wanted to understand why teenagers go nuts over rock bands. "Yes, you can study that," she said. "There is this science called psychology, which studies why people do what they do." Surprised, I replied: "You mean it's not all crazy and confusing? There's a sense to what goes on between people?" "Yes, there is." I set out on quite a voyage of discovery.

My passion has been to find out what truly heals the human spirit.

As a professional, I have come to understand psychology as intertwined with related disciplines, including: neuroscience, social sciences, evolution, mind/body health, and philosophy/spirituality. This book results from scientific study of a human's most complex organ, as well as our mind, development, evolution, body, and social nature. My goal has consistently been to develop a practical framework for organizing scientific knowledge and improvement methods/approaches for individuals and their relationships and world.

In 1996 I started writing the forerunner of this book (as some of my colleagues know) and have been both utilizing and improving the basic framework over that period. Along the way, I've found that many specialists are of compatible minds, such as Dr. Dan Siegel, generation of psychotherapists to the brain using the fist of the hand as a guide.

The evolution of this book was quite the complex process, figuring out what to include in an introduction, assimilating and accommodating the continuing knowledge explosion, complicated by changing professional and societal conditions, as well as fresh personal realizations. An article to be submitted to a peer-reviewed scientific journal is being prepared providing the line-by-line citations.

This book introduces a practical broad framework for understanding mind/brain/body/spirit, growth and development, improvement/healing, and general connections/interaction between individuals and their context. It offers a comprehensive understanding of the psychological powers and functions we subjectively experience our selves (a self) to have, grounded in macro brain-location, development, adaptation, social/habitat, and evolution patterns and dynamics.

Given that the brain-mind-self framework presents new aspects of understanding who we are – from further synthesizing and personalizing research in many disciplines, to introducing themes for categorizing and effectively working with our brain-mind-self – I wanted this initial book to cast a wide net. Through overviewing the framework, I hope health professionals, scientists, teachers, and the interested public will all find something of value.

For the interested public, encountering the powers, functions, and interrelationships among the multitude of you - as your brain/self and in your contexts – will benefit you. Fly over any scientific analysis that seems too dense. Scenarios and examples for practical knowledge abound.

For health professionals, this book's scientific tour of the universals we all live will nicely enrich trends in the profession. For example, the growing understanding that, with appropriate structure and preparation, the brain-mind-self can heal or adapt itself. Psychotherapy becomes grounded in more than a particular method or approach – even as psychotherapists work with their specific technology and processes. Through surveying the brain-mind-self framework, psychotherapists encounter a wide array of universal intrapersonal and interpersonal neuro-body-biological processes and dynamics that open the door to new possibilities.

For scientists, information specialists, and other professionals, new hypotheses/ premises offered as well as the broad review will likely stimulate further ideas, analysis, and research possibilities.

For teachers and students, this science journey aids understanding of neuroscience and psychology,

It has been 14 years since I announced to my husband and family that I was compelled to write this book – and estimated it would take a few years. They were very supportive and encouraging. Looking back, I wanted to – in my best estimation – assure that vital aspects of human nature revealed throughout history – including the noble, the boring, the worst, and the healing – were well-enough present and accounted for, so I pursued a more complete framework. However well or ill I've cobbled together this framework from science and practice, here it is! I welcome other's commentary. Certainly, this resulting baby is wanted, loved, and has its vital parts.

Introduction

<<0>>

For years, Mario's friend Kelly, a pilot with his own plane, often took him up to play in the clouds. On one occasion, things went terribly wrong. Kelly slumped over dead at the controls, as Mario watched in horror. Realizing they would soon crash, he clumsily took over the controls. As an engineer, he did his best, but the plane was careening wildly. After all, he had never learned to fly.

Then, as he describes it later, "all of a sudden it was as if someone else took over the controls and started flying the plane as me." He called ground control, who guided him in as he landed the plane safely. On the ground later he could not explain what happened.

What happened when Mario successfully flew the plane? Understood in the brain-mind-self framework, this engineer's right brain took control. Above the brainstem, we have the right and left hemispheres of the cerebrum, which can be generalized as the "right brain" and "left brain" or "right hemisphere self" and "left hemisphere self". These sides are physically separate nerve streams with specific crossover points for communication. The right brain contains our deductive reasoning powers and our highest experiential intuitive intelligence.

Furthermore, we have lots and lots of mirror neurons that activate so we can learn to do what others are doing, just from observing them do it. During all those flights his friend piloted, Mario's right brain was absorbing information, building neural networks, and learning to fly.

<<0>>

A hospital in Croatia had to call in German-speaking doctors to assist in communicating with a 13-year-old girl who woke up from a 24-hour coma speaking fluent German. According to her parents, the Croatian girl had just begun studying German at school and had been working hard, watching German television, and struggling to read German books. Since waking up from the coma, she speaks German at a much higher level than before – practically fluent – and has been unable to speak Croatian, her native-tongue.

Science is young in its understanding of how the brain acquires multiple languages, but there is some evidence that primary and later languages code differently and in separate parts of the brain.

It is possible that in-depth study of this teen may reveal some kind of brain damage to the primary language area. Nonetheless, considering our brain-mind-self neuroplastic powers, is this seeming miraculous change the work of the brain-mind-self directing its own neuroplasticity (brain changing abilities)? In combination with the functions of self-directed intention (will), behavioral practice, exposure, and neuroplasticity, did her brain-mind-self 'decide' to facilitate the mastery of German by dropping her into a 24-hour coma and replacing Croatian with German in the procedural memory networks where we find primary language memory?

What if you could choose to guide major self-improvement, repair, or recovery – without going into a coma? Understanding the brain-mind-self framework takes us closer to such power.

<<0>>

The above examples demonstrate, the extreme complexity of the human brain and being. We have a large and varied repertoire of self-capacities and functions. However, despite all our possible differences and division of labor, science increasingly understands that many aspects of our selves are universal – that all people develop and function with the same fundamentals of brain, mind, and self.

We would all benefit from a broad overview of the universally large and varied repertoire of self-capacities and functions. Pre-prepared response repertoires, including functions, capacities, and self adaptations of brain, mind, body, and spirit structurally and dynamically comprise our integrated whole.

In introducing my brain-mind-self framework, this book tells a story of us from the perspective of our individual self – the story of how mind and brain science thus far understands us, based on my hunting, gathering, and perusing of scientific works. Science teaches us that our stories and narratives are composed by the left-hemisphere. Hopefully my wiser right-hemisphere is co-guiding this endeavor.

One

Overview of Basic Concepts

As I've studied and practiced mind science (psychology), brain science (neuroscience), and their integrations, I have always thought of you and me – the individual – who has to live whatever the science talks about while needing each other and our habitat. Here are some basic concepts that flow through the matters at hand.

1. Our whole – brains, minds, bodies – functions as a **self**.

2. Each of us implicitly (something inside) knows our full **developmental** potential.

3. From the start, we grow, respond, learn, adapt, and change with an **experiential**, associative system of processing information. Later in development, we add a logical/analytical one.

4. We have a multitude of components and **parts of self** that each can have a point of view – or mind – of their own.

5. Our whole organism/self is sustained in social, cultural, family, and other habitat contexts (all of which I call '**outer**'). We have a vital attachment/social nature.

6. Humans' amazing **neuroplasticity** (brain changing ability) and our high-degree of self **adaptability** within one lifetime are related phenomena.

7. Given good-enough information, social circumstances, and other habitat systems, individuals accomplish a deep/highest wisdom, guidance, love, common-good,

sensibility, and spirit as well as have the adaptive capacity to move toward health, happiness, and well-being.

Throughout this book, I shift between thinking of a person – a self – as having different roles:

* We are each drivers, agents, those in charge who have to navigate and drive our whole being safely and successfully.

* We are the powers and functions of our being.

* We are both providers (a co-mechanic) and recipients (the vehicle) of care, maintenance, and repair.

As you go, you will find varied use of pronouns in this book. I may talk to you or about you, me, us, we, or them. It is really all you, me, and us.

More For Professionals:*

Before going into the framework, I first provide some information on the scientific grounding of concepts I use in this book, as listed above. My brain-mind-self framework is based in the premise that we are a "**self**", universally defined in the psychological field as a "basic reference to personal identity, being and experience" and the "totality of the individual, consisting of all characteristics, conscious and unconscious, mental and physical" (American Psychological Association's Dictionary Of Psychology, 2008).

Defining the individual person as a self fits with J. Scott Jordan's 2008 Wild Self Theory (WST) in cognitive neuroscience. WST integrates issues and arguments of various scientific camps in that

---

* The following section is geared more towards scientists and mental health professionals, though the public may also find it interesting, if somewhat technical.

field and avoids dichotomies such as mental-physical, mind-body, and mind-brain. On the one hand, WST validates some aspects of my brain-mind-self framework, such as grounding self in evolution and understanding inner as about outer. On the other hand, from my preliminary readings, WST is incomplete as it applies to humans and other more advanced mammalian species whose young attach to beings. WST is missing the inner key lower-upper construction method appearing with the evolution of mammals from asocial to social, which is a psychologically important demarcating juncture in human evolution

Within mind science (psychology) and brain science (neuroscience), and their integrations, other scientists/practitioners and I are still deciphering definitions along with the relationship between mind and brain. We know the brain is crucial to activities of mental life, which are subjectively experienced as mind. There are common uses like 'mind dominating mood', 'mind over body', 'states of mind'. In the brain-mind-self framework, mind includes upper.

The reporting of a newly discovered broad pattern in human psychological functioning caught my attention in 1996. Titled "The Unbearable Automaticity of Being", the authors thought the findings were hard for the individual ego to bear. We live 'automatics of being' (my term): multitudes of non-conscious, **pre-prepared response repertoires**. Circumstances can trigger such responses even before conscious left-hemisphere "will" areas of the brain have had time to assess the situation. Ideally, "these processes are in our service and best interests... 'mental butlers' who know our tendencies and preferences so well that they anticipate and take care of them for us, without having to be asked" (Bargh and Chartrand, 1996).

Watkins and Watkins (1997), hypnosis practitioners, proposed that 'automatics of being' – each with a point of view and pursuing an agenda – could be talked with as **parts of self** or 'ego states'.

With Bargh and Chartrand's discovery, mind science began to document the parts of self that dynamic and hypnosis psychotherapies had already been accessing. Working with ego states is quite widespread in many kinds of psychotherapy implicitly focusing on individuals as a self. With parts of self understood through a brain-mind-self perspective and knowledge of the experiential system, it is easier for 'conscious us' to communicate with, guide, and successfully meet the needs of 'the rest of us'.

My **developmental** understanding spans broad evolutionary biology, even as it is person-centered. Surprisingly, evidence supports my adult self development work from the strangest of scientific quarters: Blumberg's 2009 review of prenatal and postnatal animal development anomalies – 'freaks' as he calls them – reveals that each individual organism is developmentally calibrated to match "form and behavior to the challenges it is likely to face", reflecting current needs. The logic of internal developmental processes conveys "a sense of integrated wholeness, of design, of cooperation among its parts" (Dove quoted in Blumberg, 2009).

Comprising an integrated whole, our brain-mind-self implicitly knows our full developmental potential, built from the array of evolution- and development-based dynamics, functions, and patterns introduced throughout this book. Something inside knows who we are and our potential to develop/repair, as well as has the adaptive capacity to move toward health, happiness, and well-being. Given good-enough information, social circumstances, and habitats, we (except for permanent damage by physical trauma, toxic substances, or other biological harm/deficits) make our way to successful functioning via adaptability, including adult **neuroplasticity** (brain changing ability).

Our whole organism is a self – sustained in social, cultural, family, and other habitat contexts (all of which I call '**outer**'). Focused on the individual, the National Academies of Practice reviewed the latest scientific evidence and adopted a new definition of "health" in 2010: "Health is an individual's state of well-being based on integration of biological, psychological, and social functioning within the context of social, cultural, family, and other environmental conditions." This update clarifies the 1948 World Health Organization (WHO) definition: "Health is a state of complete physical, mental, and social well being and not merely the absence of disease or infirmity." Health happens in a sustainable and resilient context.

Two

Broad Quasi-Locations of the Brain

There are many scientifically useful ways to present and divide up brain function. Different ways of thinking about the brain are useful depending on the specific goals and context. The below graphic depicts a typical image of the outside of the brain.

Image 1

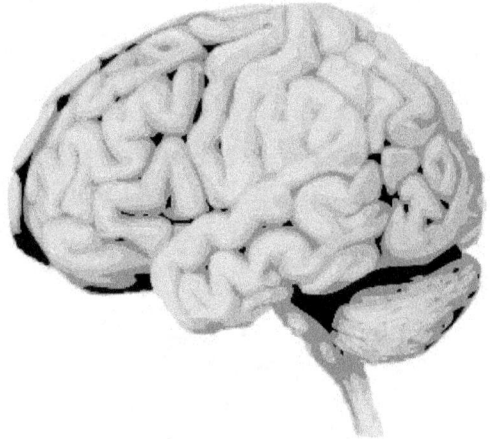

Most of what you see, the large bulge with many folds, is the cortex. However, having a conceptual framework that considers the brain roughly equivalent to the cortex is not the most fruitful in terms of you exercising your powers for accomplishing and sustaining health, happiness, and success. Because my goal is for you to be better able to manage and improve yourself, I present the brain beginning with divisions that enable you to have conduct and coordinate your roles of driver, car, and co-mechanic.

This book classifies brain components based on broad quasi-locations. I say "quasi-locations" because such classifications are largely based on location on the embryonic neural tube, with a few other considerations that will be clear as the brain-mind-self framework unfolds throughout these pages.

I distinguish the following broad brain quasi-locations:

∗   Upper/Lower

∗   Left/Right

∗   The Left and Right Cortices (which make up most of upper)

Specialized roles of front of the cortex and the front of that front (the prefrontal cortex) are also highlighted. Additionally, there are areas of the brain I refer to as the non-cortical upper. They include the cerebellum and newest-in-evolution modifications in the brainstem, which enable communication with the cortex.

This chapter briefly introduces differentiating the brain based on upper/lower, left/right, and then elaborates on cortex left/right distinctions.

Upper/Lower

Looking at the brain cut in half front-to-back, we can see the brainstem (which is organized circular not sided) and small nuclei (which are left and right pairs) in the middle, hidden by the cortex in the previous graphic.

Image 2

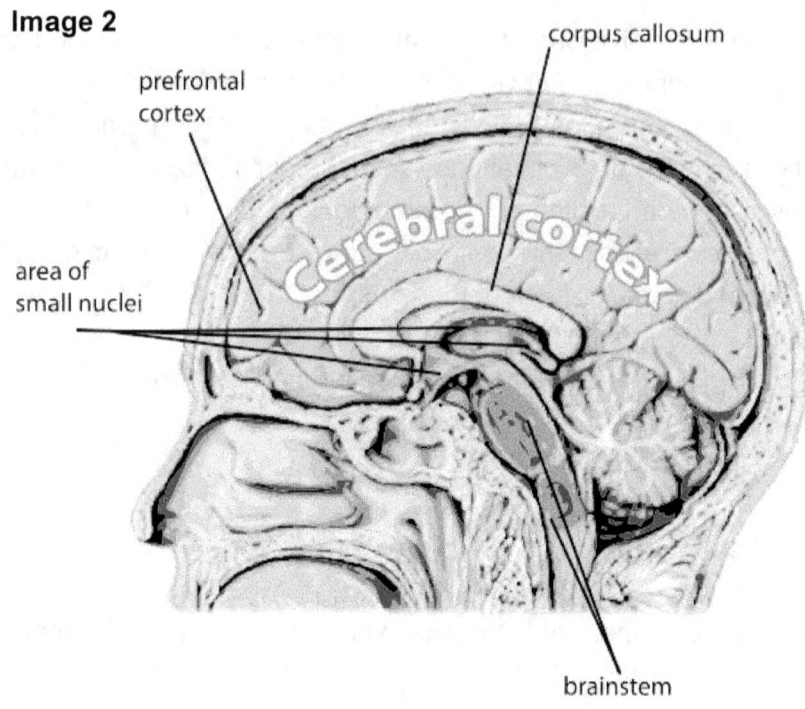

Image 3

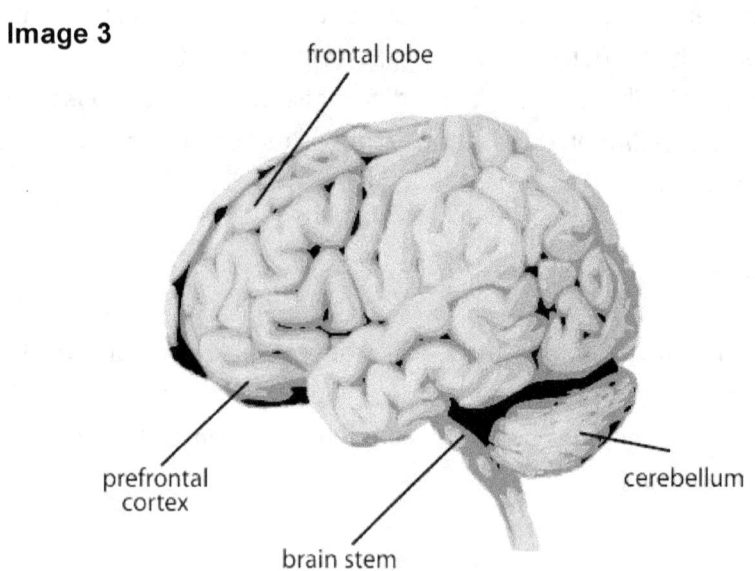

In the above inside view, we see the cortex, which comprises most of the upper, and the brainstem and basal ganglia, which function as mostly lower. A few small regions (called **nuclei**) deep in the cortex just above the brainstem, at least some of which were formed *in utero* from a mixture of brainstem and cortical tissues, function upper, lower, or both. The hypothalamus and amygdala are lower structures. The hippocampus and thalamus so far seem to function both upper and lower. That bulge out the back of the brainstem, the cerebellum, functions upper.

Upper and lower are classifications taken from location on the embryonic neural tube. What scientists call top-down brain processes tend to function upper, while bottom-up processes tend to function lower.

For the scientist's information, those small **nuclei** are sometimes referred to as the limbic system and (controversially) as a middle brain tier; however, my understanding of brain-mind-self does not use that separate middle classification, as I find it most practical – from the point of view of self-development, -care and -improvement – to quasi-locate structures upper/lower.

Right/Left

Image 3 of the outside of the brain only showed the left side. If we turn the brain so the front is facing out, we see that the cortex is divided bilaterally into right and left sides, or hemispheres. There are very few crossover points, the main one being the corpus callosum (see above graphic). The inner smaller nuclei also have a left and right version. All of these 'sided' regions are collectively called the cerebrum, which is everything above the unitary brainstem. Together, all the left-sided cerebral components are called the left brain or hemisphere, and all the right-sided components are called the right hemisphere or brain. I further discuss brain sidedness in this chapter by focusing on the cortex, with more complete descriptions – including other cerebrum structures – appearing in future chapters.

The brain made even more 'self sense' to me when I started to understand it from an evolutionary and developmental perspective. Further, neuroscience traditionally teaches psychology students about the brain and all its detailed regions and roles as though the brain was basically not about a person. Studying the brain only in that way limits the usefulness of relating brain to the practice of psychology. My framework thinks about the brain as though it is a self, and proceeds to show individuals their tremendous self-powers, begun in evolution and development, and how they can repair and improve to achieve their healthiest self.

Now that I've introduced upper/lower and left/right brain quasi-locations, and you have a visual sense of where I'm referring to in the brain, the chapters brain-mind-self framework will elaborate on the psychological significance of such divisions and how understanding them will aid in developing, caring for, and improving yourself.

Below I briefly detail upper/lower – including the common psychological skills, functions, and capacities most associated with such broad brain quasi-locations, followed by the same for left & right, then left & right cortex, along with some simple examples. Remember, you will learn a lot more about such topics as emotion, motivation, self-management, upper copies of body components, sense-perception, and motor-action in future chapters.

Table 1: Upper/Lower Functions and Capacities

<u>Upper</u>
Mostly the Cortex

- Thinking, Cognitive Function, Ideas, Imagination, Language
- Representational Ability, Symbolic Capacity, Theory of Mind
- Beliefs, Moral Ideas/Values, Humanistic and Social/Caring Sensibilities
- Planning, Reason, Will, Conscious Decision-Making
- "We" Intentionality
- Concepts, Frameworks, Constructs, Theories making sense of inner & outer information, events, experiences, narratives
- Causal Belief, Interpreter
- Effective Procedures for Performance, Success
- Mirror Neurons (most)
- Conscious Awareness/Attention, Observing (all outer & inner)
- Sense of Selfhood
- Self Awareness, Mindfulness, Mindsight, Meditation
- Copies of Body Components
- Upper Memories (conscious and/or unconscious); storage space
- Sense-Perception & Motor-Action Integration Functions
- Emotion Integration Functions
- Motivation (instincts and urges) Integration Functions
- Self Management & Control (scientists' term: self-regulation)
- Observational and Vicarious Learning
- Learning by Teaching
- Increased Habitat Flexibility and Adaptability
- Final Attention and Performance Self Stage Managers (cerebellum/new brainstem)

<u>Lower</u>
Mostly Subcortex (Including the Brainstem) and Body

- Emotions
- Instincts, Basic Motivations/Urges (stay alive, food/water, air, sex, home territory), Compete/run-off/kill what is interfering/ competing
- Biological Rhythms (such as seek, work, acquire, consume, rest, sleep, play, reward, pleasure/pain, time)
- Lower Memories (many unconscious)
- Vagal Nerves
- Physical/Body Needs and Functioning (e.g. adrenaline, speech patterns, indigestion, nervous energy/movement, blood pressure)

Given the operating dynamics of the whole organism as brain-mind-self, the **body** – all its organs and tissues, nerve endings and individual cells – constitutes part of lower. This book and the brain-mind-self framework use "lower" to mean both the lower brain and the body. To give one example of how the body counts as lower, take the "stomach brain" – a fairly large and dense collection of nerves that can have its own reaction to stimuli, such as flooding the stomach with acid (thus keeping antacid companies in business).

Table 2: Cats Example

Consider what happens when you read the word "cats".

Upper

- What facts do you know about them?
- What are your ideas about cats?
- What do you believe about cats?
- What images or fantasy come to mind?

Your answers to these questions activate upper brain functions.

Lower

- What emotional associations do you have to cats?
- Are your feelings negative, positive, or a combination?
- Do you have urges to avoid or approach cats (motivations)?
- What body sensations do you notice?

Your answers to these questions activate lower brain functions.

Your lower activations about cats also are processed (internally worked on) in upper. We integrate our emotional reactions and instinctual urges with upper capacities and sensibilities as well as care for, control, and manage self.

Noticing your internal reactions to this exercise is using your upper capacity sometimes called observing, mindfulness, mindsight, awareness, or conscious attention.

Lower and upper operate both rather separately and as an integrated whole. One practical analogy is that of a **two-story house**: lower is the more-basic first story, and upper has left and right sides that can sometimes function as though they were removed from lower. Actually, the left is better able to separate, as the right is postnatally grows more extensively integrated with lower, which will be further discussed in Chapter Seven.

Finally – for any biologist readers – biological science generally refers to lower (subcortical circuits) as reflexive motivational and emotional activations, which need the non-reflexive cortical (upper) capacities and constructions to negotiate their environments adaptively and guided/controlled. Without that inner guiding regulation and communication from upper, lower visceral (gut), motivational, or emotional circuits sometimes react in ways that 'conscious us' (upper) does not understand or want (including what are called 'low road' reactions).

Table 3: Common Upper/Cortex Sided Psychological Processes

(**Bold** points directly correspond to something in the other cortex.)

Right Cortex	Left Cortex
•**Deductive Reasoning/Logic** •Imagination, Visualization Images, Fantasy, Artistic Ability •**Non-Verbal Language** (tone, accent, context, pragmatic) & **Rudimentary Language Processing** •Facial Perception •Spatial Manipulation •**Approximate Number Computation and Estimation** •Sense-Perception and Motor-Action Integration Functions •**Extremes Of Emotion Processing** (both positive and negative) •Intelligence for Emergency, Chaos, and Novel Situations •Innate Sense of Time and Space Orientation •**Sense of Selfhood** •Experiential Associative Thinking and Knowing •**Conscious Awareness Integrated with Body and Emotions** •Sensory Processing/Mixing •Highest Level Of Wisdom and Guidance, in Touch with What Can Be Experienced as a Higher Power •**Personal Felt Memories** •**Long Term Memories**	• **Linear Reasoning, Inductive Logic, Detail Focus** • **Verbal Language, Naming, Labeling** • **Alphabet, Reading Language, Writing** • **Math, Counting, Arithmetic** • Thinking/Reasoning about Ideas • **Middle Range of Emotions Processing** (not extremely positive or negative) • Sense of Will or Agency, Choice, Pushing-Oneself • **Social Self-Presentation** • Interpreter • **Conscious Awareness Experienced as Separate from Body and Emotions** • Coherent Narrative Discourse • **Word and Spoken Memories** • **Fact-like Memories**

Here are some further descriptions or examples of these left and right differences in thinking:

Deductive reasoning, which relies on premises we already know are true, is distinct from inductive reasoning – drawing conclusions that are probably true, such as based on observation. Psychological experiments show the former happens in the right brain and the latter happens in the left. For example, when you realize solutions in your sleep, your right cortex uses information you already know to deduce patterns and conclusions; whereas making conclusions about the world based on previous experiences (such as that this new cup will, too, hold water) counts as inductive reasoning done in the left cortex.

Sensory processing mixes information received from the various senses and further processes that mixture with all your other cortical capacities as well as any additional internal knowledge about you and your circumstances. In application, the right hemisphere can strategically reinterpret external sensory input without the left knowing there was any internally generated change.

Extremes of emotion are processed on the right side, midrange emotion on the left. To illustrate, when two friends run into each other, one will often ask the question: "How are you today?" If the question elicits strong negative or positive emotions, you are processing that question in the right cortex. If the question, elicits mild feelings (such as a typical "Fine, thanks." response), you are processing it on the left.

Crunch! You just stepped on something hard, but with a little give. Your right brain will integrate your sensory input and decide if what you stepped on was a snake or a stick. Any influence of context, personal felt memories, your highest experiential-intuitive wisdom and guidance, fantasy, imagery, activated extreme emotions, and deductive reasoning also happens on the right side. Influences of will power, social presentation, fact memory, internal conscious awareness (separate from body and emotions), linear and logical processing, and thinking about ideas all come from the left side.

Together the left- and right- frontal cortices (actually the pre-frontal cortex; pfc) plan, decide, and think through a topic/issue, when necessary.

Three

Self Evolution, Development, and Adaptation Themes

We can take fuller advantage of our brain, mind, and self powers and functioning when we understand and integrate key themes in human evolution and development – as well as working with the whole as a self – into the already accomplished arenas of psychology, psychotherapy, and psychology-neuroscience collaboration.

Looking down our phylogenetic line, we observe a land-vertebrate reptilian-like common ancestor with which we share the lowest parts of our brain. This chapter explains not just how we got from that ancestor to our modern human brain, but also the psychological significance of specific changes.

Studying the long pattern of phylogenetic evolution and individual development, two important themes increase our understanding of brain-mind-self: asocial-social-asocial responsiveness and lower-upper construction.

These themes are inherently connected. Our individual self results from a lower-upper redesign, evolutionarily built starting with the mostly-individual (as opposed to social) nature of the reptile-like ancestor common to all social mammals, and complete by the time of *Homo sapiens*. Thus, the self was fully formed before the 'culture of modern humans' emerged (explained in Chapter Six).

The significance of such themes lies in their useful categorization of psychological components everyone has, where patterns often go unrecognized. Understanding aggregate groups of self functions and capacities allows the individual to manage these components with the aim of optimizing health and well-being.

Theme One: Asocial-Social-Asocial Responsiveness

Neuroscience has long identified that "instinctual, nonsocial operating systems lower in the [human] brain", which are responsive to habitat conditions, are in humans "filtered, regulated, modified and used in construction by the vast layers of higher cognitive, emotional, attachment, and social complexity that learning and neuroplasticity permit" (Panksepp, 1998).

Our brain-mind-self has an early mode of responding asocially to its environment that interacts with everything – even people – instinctually, instrumentally, or as though they were things/objects. On top of that, we have social/humanized responses to environment that integrate with and manage the previous asocial components, such as needing and valuing personal identity, people, relationships (attachment), and groups.   Finally, our latest-in-evolution asocial/impersonal responsiveness to environment involves our capacities for discerning cause and effect, scientific experimentation, and technological innovation.

This asocial-social-asocial pattern manifests both in the phylogenetic evolution from reptiles to modern humans and in the individual development of each self from infancy through adulthood. All three components continue to operate in the normal adult self.

Note, particularly, the normalcy and importance of lower asocial. Comparing it with the reptilian brain frequently leads people to think of that part as bad, mean, or about other negative associations they may have to reptiles.   In fact, we have the human version of a reptile's brain, and it is a natural and necessary part of ourselves.

My experience is that people do not consciously realize the wide range of ways early-in-evolution asocial aspects flow and integrate with other facets of themselves – indeed with the whole brain-mind-self.   Individuals and groups can care for, structure, and manage this lower asocial to both enjoy the positives of life for, and minimize the harm from excess or deprivation to, the individual, loved ones, endeavors, civilization, and habitat.   Such asocial satisfactions can come from: work, rest, food, sex, seeking (payoffs), and everything else on the Lower Asocial part of the chart in Chapter Two. Additionally, our lower asocial includes the motivation to feel secure from the threat of loss or deprivation – on its own it seeks to hoard our supplies, and will do so unless upper/outer intervenes.   Lower asocial agendas are the components of any individual or group-organized endeavor that focuses on money payoff, for example.

Many areas of our self start out with lower asocial responses and have to be brought up into our higher social level in order for us to have a "filter" or degree of control on our behavior based on social/humanized contexts.   Integrating the asocial into our social responsiveness means that social patterns and rules can influence behavior previously governed only by our lower thing-, object-, or lizard-like brain.   This 'individualizing', 'personalizing', and 'socializing' changes the way the brain organizes information and adds felt and believed constructs/values/cognitions like rights, roles, respect, and the social good.   Among other benefits, this process enables us to work in teams and control urges that would be perceived as too aggressive, violent, selfish, or offensive.

In addition, the middle social/humanistic aspect includes the human ability to personify parts of self – to understand our selves as not just comprised of functions and instincts – and as not just objects, but as having "person-parts" with their own needs and points of view (this concept is currently often called ego states).

Because of our hierarchical asocial-social-asocial pattern, humans experience parts of self either as an object- or thing-like "it" (using the first/lowest asocial) or as a person-part having individual needs/perspectives (using the higher social).   Both are true, but one may be a more useful framework for a particular person in a specific situation and time. Depending on the context, an "it" part can be seen as an inanimate object like a computer component – e.g. viewing depression in terms of a chemical imbalance – or as instinctual/physical (a biological part of us, but automatic or unconscious) – e.g. having physical needs like sleep or help to calm down.

The capability to possess personal/social identities set the stage for humans having values – a sense of fairness, loyalty, belonging, and what is right and wrong. Social normally and necessarily manages earlier-asocial components. Chapters Six, Seven, and Eight elaborate on all these social influences.

Higher evolved young mammals developed longer and more helpless stages of growth that paralleled the transformation of solo-surviving species with instinctual/hormonal maternal care systems to species that include attachment, group surviving, and reproducing relationships.

So far this chapter discussed the earlier/lower asocial, but barely mentioned the later/higher asocial. The specifics of this latest-in-evolution asocial/impersonal responsiveness to environment are discussed in Chapters Six, Seven, Eight, and Eleven. Here I mention my proposition that the unique social and personalized organization of humans was the context that enabled the evolution of the latest asocial intelligence and individual will/control capacities.

An upside of our latest-in-evolution higher asocial is that it gives individuals our capacities to improve and regulate (direct, control, readapt, or reprocess) the rest of self and social.

Our higher asocial also poses downsides to society. In conversation in 1996, a genetic bio-ethicist calmly described to me that duplicity (deceitfulness) is an alternative adaptation strategy available to humans, should they be able to completely not care about their loves, loyalties, bonds, and the common good. Over my years of study and practice, I began to understand how this strategy was possible.   Eventually, our changes in habitat and group dynamics enabled humans' newest-in-evolution capacities to include the potential for an adaptation strategy where we could mentally step out of the bonds themselves and live in duplicitous, hidden, or pretending ways. Instead of forming real bonds, people employing this strategy may have pretend connections, and live rather phony, internally isolated, or criminal lives.

While this first theme was relatively simple to initially describe, the second requires a more detailed explanation.   Its previous mention in Chapter Two provides a basis for the substantial discussion below.

Theme Two: Lower-Upper Construction

I propose that the brain-mind-self added a lower-upper neurological construction method through evolution that it employs in developmental growth (beginning near-birth) and current adaptation (experiential learning, memory, and neuroplastic re/adaptation).   Identifying brain communication and information processing circuits as "bottom-up" or "top-down" is already part of some scientific conceptualizations.   The brain-mind-self framework adds explicit articulation of the general lower-upper construction methodology and operating process.

As mentioned in Chapter Two, lower refers to most of the brainstem, a few structures just above it, and the body; while upper refers to the cortex and some regions below it.

The following six subsections (A-F) describe lower-upper construction in evolution, development, and adaptation.

A. Evolutionary Origins of Lower-Upper Construction

Reptilian or 'lizard-like' species have their sense-perception (see, hear, touch, taste, smell) and motor-action (walk, swim, eat, drag, turn, blink, etc.) brain functions on the same level as the rest of their brain. This single-level entity's ability to sense and act is effectively instinctual and asocial. In humans, the brainstem and its behavior-coordinating basal ganglia retain this reptilian same-level unit, and are part of what I refer to as lower.

Additionally, over the epoch time of species' adaptations in our specific phylogenetic line, a large part of the sense-perception and motor-action functions moved up into the central and back areas of the left and right cortex, as Chapter Five describes.

Further, I postulate that lower experiences of interacting with the habitat (part of 'outer') are built into upper during construction. This new method of learning – building experiences of outer into our very construction – greatly adds to an organism's ability to remember, have memories, and adapt, as mentioned throughout the book, especially in Chapter Ten and Section E below.

From where do I get the terms 'upper' and 'lower'? They are not strictly physical locations in the adult human brain. Gestation and what humans have in common with reptiles illuminates the use of these terms.   Called the neural tube in early gestation, the structures we share with reptiles are the brain areas common to all vertebrates. At the tube's top end, primitive mammals evolved right and left lobes of several nuclei forming a kind of circle around the brainstem. In humans, this circle appears on the bottom of the cerebral hemispheres as the limbic system, the first part of upper. All the other upper structures of the human (and other higher mammalian) brain also develop as endbrains off the neural tube, which itself becomes lower. During fetal development, parts of the brain fold over on themselves to become the final structure. Hence mammals do not have brains that divid geographically into upper and lower with a clear line, but if you look at their brains in early gestation, the neural tube is lower and everything growing off the top end is upper. Evolution follows a similar pattern. While 'upper' and 'lower' are useful designations with evolutionary and developmental significance, they can at best be called 'quasi-locations' in child and adult brains.

B. Depictions of Lower and Outer in Upper

A Psychology Laboratory Scene:

The mouse smells the cat and scampers away. The chase is over almost as soon as it begins, as the mouse darts into the tiny corridor. Without even a pause, it scoots through the barely visible crack in the wall. This mouse is hooked up to electrodes, which show how it is following the previously constructed and practiced/memorized habitat-body map in its hippocampus in order to so precisely and quickly escape.

In order to quickly turn and run through the hole, the mouse not only follows the habitat, but also a representation of the habitat in its hippocampus, part of upper. This self-constructed map is based on the mouse's individual experiences and includes representations of its body. The mouse experienced particular aspects of its habitat in association with specific parts and functions of its body. These habitat-body associations are neurologically connected in the mouse's hippocampus. In humans, multiple places in the brain – such as the hippocampus and the sensory and motor cortices – routinely form such upper representations (also called maps, templates, or grids). In short, most mammals map experiences of habitat. This mapping includes the experiences associated with body, arranged somatopically (see Chapter Five). Furthermore, this mapping includes tracking significant resources and dangers/problems in the habitat.

Understood in the brain-mind-self framework, higher mammals (in contrast to same-level reptiles), have both lower and upper response involvements with their habitat and body. Physical interaction with the external world for higher mammals involves both the body and upper internal representations/mappings.

Do to its internal mapping in upper that can now influence lower, the mouse has a greatly increased ability to experientially learn and relearn in formative and adult years. This flexibility applies to other mammals as well, including humans. Sometimes neurons stop firing after prolonged exposure to a stimulus – think of someone no longer startling at a regularly scheduled train, periodic machine beep, or habitual animal noise. Such sensory-motor adaptation effects are examples of the human brain's upper communicating with lower to override the shock response that would otherwise occur.

In *Homo sapiens*' environment of evolutionary adaptation (EEA; explicated in Chapter Five), the brain learned/mapped the location of potential and actual food, water, escape routes, predators, mates, offspring, social groups, safe places, and home territory. Learning to safely climb trees/ladders, operate tools and vehicles, explore natural formations, or play sports/(video)games is all part of adapting to one's physical habitat in the modern world. The brains of addicts map the sources of the addictive element as well as the paraphernalia to feed their habits, whether for retrieving/ingesting drugs, compulsive sexual or eating behavior, and so forth. (Chapter Four discusses addiction in greater detail.)

Mental representation or dual coding theories/evidence (e.g. Paivio, 1990) fit here: perception and response not only involve external information, they are dual coded with what that external information means to the self.

C. Lower-Upper Self Growth

Lower-upper construction methods operate in the brain's self-growth and development from near-birth through the completion of the formative years. Social, emotional, and self-developmental aspects of individuals grow utilizing lower-upper construction methods.

The human brain near birth essentially functions with the brainstem plus a few other structures (notably the 'emotional' amygdala and the hypothalamus). While the basic brain tissue of the upper – i.e. sensory, motor, emotional, and executive functions, as well as their integration – are prenatally built, they must mature postnatally to 'come online' (via lower-upper construction).

Further, an individual's upper must be constructed partly based on lower's experience of outer. Growing an upper occurs in species where adults survive and reproduce in families (interdependent groups that bond within themselves), such as many higher mammals including humans and other primates. Like the mouse from the example in section B, individuals of these species create internal representations of outer. More than just mapping physical location, however, these copies include associations the individual (i.e. lower) has to the original. For example, young begin to map their experience of who/where/what is protecting, nurturing, teaching needed adult skills, or otherwise helping – and conversely, who/where/what is potentially harming, hurting, or otherwise making things harder.

In order to grow an upper, you need others (often adults) to protect and take care of you. Longer childhoods resulted from requiring more time and experience to grow this upper. Upper construction may have created a positive feedback loop with species turning social because of this increasing need to team up with others. Emotions became a signal system between lower and upper, as well as self and other. Species got more cognitive abilities: to think, plan, strategize, and – most importantly – have experiences become internalized. This new upper provided space to store memory – and not just 'thing' memory, but associative or linked memory (e.g. here's what happened and here's what I was doing when that happened and how I felt about it). Hence lower's experience of outer participates in the construction of upper.

D. Lower-Upper as Self and Parts of Self

More than biological developmental dynamics and functioning, lower and upper can also be understood and interacted with as certain roles or parts of a person – of your self.

Upper and lower can be types of 'automatics of being' – self parts that have a point of view and goal/agenda. Lower includes parts that need (signaling their need with emotion, distress, or body symptoms) and that ground or center the self. Sometimes these lower components can be classified as child parts, but they also encompass normal adult physical, social, and caring/nurturance needs and grounding – i.e., emotions, motivations, and instincts. Lower self needs are met by upper and/or outer (other people and the rest of the environment).

E. Lower-Upper Self Adaptation

Humans evolved certain universal structures that enable us to successfully live and cope in varied environments within one lifetime. Such an evolved capacity is a species adaptation. However, evolution and some areas of psychology have different definitions for the word 'adaptation'. As opposed to the macro scale of evolution, numerous sub-disciplines of psychology and psychotherapy use 'adaptive' and 'adaptations' to describe changes in the individual within one lifetime, i.e. learned adaptations. Theories and research conceptualize such adaptation – e.g. by J. Piaget in cognitive developmental psychology, by M. Gazzaniga and D. Siegel integrating neuroscience and mind science, and in information processing works such as that of S. Epstein and F. Shapiro.

Here I am proposing that lower-upper configurations, constructed and modified as we cope with our varied environments, be called adaptations. Our brain-mind-self makes self adaptations. I add the word self when it seems useful for the individual to experience or internally understand adaptations as parts of self.

We neurologically code copies of people, places, and events in our cortex via lower-upper construction. In doing so, a part of us automatically makes conclusions (beliefs, generalizations) about our self, others' selves, people in general, groups, the world, and their interrelationships – including about reality, identity, and existence. A part of us just happens to get this neural-coding job, and that part then instantaneously shifts into living as if it is who/what it has just copied.

The brain-mind-self framework's understanding of adaptation fits with Jean Piaget's developmental psychology theory of cognitive adaptation, a constructivist theory of human learning and instruction about the properties of the physical world. In Piaget's theory, the interaction of experience and cortical growth enables the infant to develop schemes or cognitive constructs about the physical world around them. Comparably, in the brain-mind-self framework, the human individual develops self and social constructs/beliefs/identities from the interaction of experience and cortical (re)growth. Self adaptations, using the lower-upper construction method, go further than constructs and rules, however, in that the brain makes actual copies and lower's experience of outer is coded internally in upper.

F. Self Adaptation To Any Significant Encounter

One kind of lower-upper construction occurs in initial development. As a newborn, we function mostly lower (mostly brainstem). Throughout childhood, the developing self acquires more and more upper capacities and integrated functioning. In initial near-natal growth, lower's experience of outer shapes upper's development and structure. As upper structures become operative, they have their own involvement with lower. Future chapters elaborate on this brief summary of child self development.

Here I introduce the premise that our brain-mind-self continues throughout life to use the lower-upper neural construction method in forming self adaptations. Whenever our self assesses that we had a significant experience, we not only remember it, we neurologically make an internal copy of the whole thing. Whatever of your capacities and functions were operational, activated, and participating at the time (such as your fear circuit) and whatever was done to you (let's say you were attacked by a family member) become a part of your self that reenacts the experience every time it activates. The part of you whose job it was to make the copy gets neurologically 'topped' by its creation, thus becoming a functionally lower neural network. I'm adding here the concept of 'functionally lower' and 'functionally upper' neural networks, where lower and upper no longer reference quasi-brain locations at all, and instead refer to construction roles. Any combination of parts of the brain – comprising lower, upper, or both – can form its own copy of an event, where the copy now functions as upper and the original part functions as lower to that copy. It is as though the new functionally lower part wears a neurological mask/costume, runs a movie scenario, or otherwise activates the representation of the whole outer event/persons that function as its neurological upper. Using the movie projector analogy, the movie plays in your head because lower sends up the neurological life to power it – with lower being whoever you were at the time you made this self adaptation.

You can think of the new upper copy as resulting from a costumed/masked lower – and not just a scarecrow or doll representation type, but actually an animated presence of lower's perception of outer. Lower you wears a costume/mask, while not subjectively knowing or experiencing that s/he is doing so.

Frequently in psychotherapy, I work with lowers that are child parts of one's self needing to be nurtured, protected, and loved. A self adaptation's functional lower is whatever age the triggering event occurred.

While lower created and powers upper, it is in fact upper that holds neurological dominance over lower. Lower cedes its power to upper because the whole determines that this self adaptation is helpful. However, in the context of a therapeutic repair with some outer involvement, the brain-mind-self may decide that such a self adaptation is no longer necessary or overall harmful/maladaptive. In this case, lower can take its power back and, if healthier options exist, move to an upper that better meets its needs – essentially reprocessing the no-longer-needed upper adaptation so the neurons to be used elsewhere.

Virtually all of this learning and lower-upper construction is done with our experiential system of information processing, as Chapter Nine explains.

Four

More About the Lower

Work, movement, rest, play, sleep, dream.
Air, home territory, food, water, sex/mating, material goods.
Approach, avoid.
Fight, flee, freeze, faint, befriend.
Live, die, kill.
Predatory, preyed upon, hated.
Sadness, loss.
Fear, panic, terror, rage.
Joy, ludic delight, passion.
Love, nurturance, safety.
Posture, gesture, body care.

These are some of the motivational and emotional well-springs that comprise our lower, and which activate based on current goals, challenges, and stresses. Individuals aspiring to high-performance or wishing to sustain/recover well-being frequently benefit from information about how such lower processes relate to their specific situation.

My goal here is to have conscious you become more familiar with what scientists have learned about the biological vicissitudes of our self capacities that begin lower (meaning they have needs from and are responsive to outer and upper). These components of our self are involved in our psychological troubles as individuals, families, societies, and as a species.

Scientists have diverse ways of describing and studying these evolutionary birthrights we've all inherited, and in the process dice and slice the lower into functions, needs, and reaction circuits.

Here is my classification of lower self functions and capacities from which I frequently draw in my work:

Table 4: Lower Self Functions and Capacities

∗ Lower Asocial/Body

Lower Asocial

1. Earlier asocial involves physical components, including: needs (survival and reproduction), rhythms, and responsiveness.
2. Asocial understood as an important part of our personal/social identity and functioning.

Lower Body

3. Lower brain structures mostly-directly run our vital life organs (i.e. gut body, visceral systems).
4. Lower includes all actual body characteristics, functions, and systems (where body is being body, not behavior).

∗ Lower Organized Socially/Emotionally

5. Asocial evolutionarily organizes into basic emotional and motivational/instinctual systems common to all mammals.
6. Our pleasure/pain and 'natural feel-good' neurochemical/hormone systems provide the neural substrates for these specific emotional and motivational processes universal to mammals.
7. An evolutionarily older wellspring of attention supplies our lower – one that is biologically alive, present, and alert.
8. Lower functions and capacities hold localized memories – including in the brainstem and body, whose are largely nonverbal.
9. Lower is the base and beginning of personal needs/identity, attachment, social, and moral/ethical dimensions of our selves.

Lower Asocial/Body

People often find it useful to understand that everyone's self has important early asocial and body aspects, which we experience without emotions. These lower basic-survival/reproduction components include:

- Physical needs – food/water, reproduction, safe home territory
- Instincts – predator avoidance, predation, aggressive stopping of hurter/competitor
- Biological rhythms – daily, monthly, or seasonal; sleep/wake; rest/work; reproductive cycles
- Goal-oriented actions – approach/avoid, orient towards/away, explore/forage/seek, work, acquire, consume, or intimidate/kill what is interfering/competing/damaging
- Body's characteristics and functions

Yes, many of the previous specifics overlap, which helps illustrate how easily people can have conflicting or compound motivations/urges.

Additionally, we have the human version of other responsive systems that vary considerably by species, such as energy (body regulation), temperature regulation (reptiles are cold-blooded; mammals are warm-blooded), timing rhythms (e.g. sleep/wake, rest/work, dreaming, calendars/cycles), acquiring and consuming behavior. To clarify, seeking (the thrill of the chase- the thrill aspect is mammalian), acquiring (possession), and consuming (utilization) can be independent of one another.

Lower brain structures mostly-directly run our visceral body organ systems, such as: digestive, respiratory, and cardiovascular – including the vital heart, lungs, kidneys, liver, and pancreas.

One biological element useful in psychotherapy that clinical work has already highlighted is the vagal nerve and its change from reptilian fight-flee to mammalian fight-flee-freeze and then, in humans, to fight-flee-freeze-befriend. Humans have twelve pairs of cranial nerves connecting the body to the brain, all of which connect to the brainstem except the first two, optical and visual, which connect directly to the cortex. The 10th cranial nerve – the vagal nerve – participates in the fight-flee reactions of all land vertebrates, including by influencing adrenalin surge or dampening.

While reptiles have a simple fight-flee reaction circuit, mammals' vagal nerves have a third branch allowing an additional response – freeze – that seems to serve two fundamental biological purposes. If you are truly becoming an animal's lion's lunch, you will feel no pain.   However, it is also a last ditch survival strategy, as periodically when the prey/victim goes limp (sometimes called "playing dead"), the predator loses interest or otherwise ceases. An infant who won't stop crying may go into a freeze mode when hit by an abusing parent; yes, the child no longer cries, but it also lives detached from its own needs and emotions while in that state.

Higher mammals have a fourth (pro-social) vagal branch that supports us being calmer and less afraid around others, which is part of our "tend and befriend" reaction.   This fourth vagal nerve becomes the default actor in socially or intellectually engaging situations and produces a state of pleasant, not overwhelming, arousal.   Bad social experiences can dampen down these good social arousal activations as a person – perhaps unwittingly – becomes conditioned to be uncomfortable with and avoid people. In humans, the vagal nerve can play a role in PTSD (Post Traumatic Stress Disorder), depression, unwanted learned avoidance, or other health problems.

Mammals

Moreover, the lower asocial provides an important part of our identity and functioning.   We need and enjoy food.   Physical experiences are a fundamental part of life, as are emotions – that quintessential form of communication between our lower and our upper, as well as others.   An infant needs to work out life rhythms based on its particular habitat. A two-year-old child needs to first develop a sense of ownership and "mine" before s/he can successfully add the sensibility to share and take turns. Possessing supplies is part of the lower asocial, but integration into social (which is largely dependent on upper capacities) determines to what and when to apply asocial components. Thus people use their lower asocial integrated with social to better determine when it is appropriate to share or take ownership of something.

My experience is that people less consciously realize the wide range of ways early-in-evolution asocial aspects flow and integrate with other facets of themselves – indeed with the whole brain-mind-self. Individuals and groups can care for, structure, and manage this lower asocial to both enjoy the positives of life on the physical level for, and minimize the harm from excess or deprivation to the individual, loved ones, endeavors, civilization, and habitat. Such asocial satisfactions can come from: work, play, rest, food, sex, seeking, and everything else on the Lower Asocial part of the chart in Chapter 2. Additionally, our lower asocial includes the motivation to feel secure from the threat of loss or deprivation.

Lower Organized Socially/Emotionally

In tandem with mammals evolving a lower-upper brain construction, the earlier asocial rearranges to become the lower of an integrated system, as opposed to a sole-functioning single-level unit in reptiles. Hence, the asocial organizes emotionally (becomes able to have emotions) along with transforming into the lower of complex social, cognitive, and sensory/motor systems. Becoming the lower of social means that an individual needs others of their own species (attachment and/or adult teams); as the lower of cognitive, the brain's learning and memory abilities dramatically increase; and what it means to be the lower of multi-level sensory and motor systems will be detailed in the next chapter. This chapter focuses on social and emotional roots in lower because those tend to be relatively less addressed in mainstream psychology. Furthermore, asocial-social integration is one of the key driving forces behind these other changes – and indeed the whole lower-upper construction method of development. For example, as mentioned in Chapter 3, social dependencies allow an individual the time and experience needed to grow a salubrious upper.

Furthermore, asocial-social integration organizes a large part of lower functioning based on specific systems. Newborns begin navigating the world with much of their asocial organized into seven emotional and motivational/instinctual systems common to all mammals:

- ♦ Fear
- ♦ Anger/rage
- ♦ Lust/love
- ♦ Grief/panic (loss, attachment separation-distress/abandonment/self-annihilation)
- ♦ Seeking (supplies such as food, drink, sex, social needs)
- ♦ Play (laughter, singing, dancing, dramatic enactments)
- ♦ Care

These mammalian motivational and emotional processes build off of our pleasure/pain and 'natural feel-good' neurochemical/hormone systems. The basic seven occur in all mammals, with details varying by species – hence we have the human version, but all mammals have their variant as well. I give examples below. I reserve the seventh, care, until the end of this chapter.

More About Lower

Additionally, we have the human version of other responsive systems that vary considerably by species, such as energy (body regulation), timing rhythms (e.g. sleep/wake, rest/work, dreaming, calendars/cycles), acquiring and consuming behavior. To clarify, seeking (the thrill of the chase), acquiring (possession), and consuming (utilization) can be independent of one another.

Some classic psychology experiments with animals illustrate lower us. Ever heard of the white rats in psychology experiments that learn to press levers for rewards? Or the dogs that become depressed when they are powerless to prevent electric shocks to their cages? Scientists set up the habitat for them, their 'contingencies of reinforcement'. Scientists showed they could change the payoff schedules and so forth in such a way as to change the behavior or mood of the animal, even to the point of getting the animal to work itself to death. These forms of associative and contingent learning are labeled classical and operant conditioning. Panksepp (1998) notes that it is our lower – our visceral, emotional, and motivational circuits – that are shaped with classical and operant conditioning. Many a gambling addict has had his/her "payoff contingencies" shaped by gambling machines. 'Hunting and gathering' video games that many young people play involve combing the territory until the treasures are acquired. If you encounter blocks, problems, barriers, or competition in the more violent kind, the player 'blows them away'- an asocial or predator kill response to what is stopping/ hurting you. Anger circuits can also be activated.

These lower functions relate to many real-world scenarios.

Example #1 – Crisis Reactions: Fear, Anger, Stop/Kill, Flight, Freeze

I roll the jeep five of us are riding in. Thankfully, no one was physically hurt. You perceive the experience as exciting. I am not just afraid, I have defecated. The third person is screaming and yelling: "You idiot; you dumb-ass! I'm going to beat the crap out of you! I'm going to kill you."

The fourth person jumps off the jeep and runs away as fast as she can.   We look over and the fifth person is sitting there completely stoned faced and numb.

You had a thrilling roller coaster ride. My emotional fear circuit fired. The fourth person enacted the flee/avoidance reaction. The fifth, the freeze circuit, so is numb and frozen in fear. Person five seems in a fog for days.

The third person is enraged. Our anger circuit is called our anger/rage reaction because it is designed to go from 0 to 100 in nano-seconds, if needed. Afterwards, person three spends days raging at me – indeed, everything makes him mad. The previous "I'm going to kill you" is the asocial urge to stop what is threatening to hurt you.

I have a greater urge to spend a lot of time with my close friends and family. This desire relates to our need for attachment and group belonging (see Chapters Seven and Eight).

Example #2 – Addictions, Extreme Payoffs

A man was so addicted to on-line porn that he would get an erection if he even approached his computer. Psychotherapists are experiencing a major increase in requests for help from people who have watched extensive amounts of video porn on the Internet and wind up with behavior they are having trouble hiding or controlling.

This man exemplifies how human nature retains 'lizard brain' motivations that are asocial and not humanized with empathetic values/sensibilities.   His sexual payoff receptors were over-stimulated and his brain learned to associate the sight of his computer with sexual stimulation, much like Pavlov's dogs salivated at the sound of a bell.

In addition to emotional and social influences on lower asocial, upper cognitive powers can also determine resulting behavior. Cortical parts of the brain enable people to consider what is smart vs. stupid for the individual, regardless of the person's values, social loyalties, and short-term opportunities. "Okay, I would like to, but is it worth the cost? Is it really viable for me to lead a double life? Would I really be happier in the long run doing...?"

Lower motivations and needs of the self, both asocial and social, are the well-spring of natural feel good hormones. Over-exposure to drug substances, gaming/gambling, high fat/sugar foods, or sexual material can lead to these hormones over-stimulating built-in reward circuits. The brain can become maladaptively conditioned to a specific thing or substance to the extent that it no longer acts in the whole's best interest, but instead focuses on achieving the associated hormonal reward. When the man in this example becomes uncontrollably responsive to stimulating cues and triggers, in this case Internet porn and even the laptop, asocial aspects have taken over his self – the whole social and person. This phenomenon is akin to when someone can be described as "married to the alcohol" or "married to their work" (see the Scrooge example below).

Furthermore, addictions and eating disorders involve instinctual urges that are damaged by habitat excess/deprivation, impaired by distorted payoff contingencies, or being used to cope with other self problems. Addictive drugs, doctored sexual images, and excess junk foods (like high fat/sugar, low fiber) are all examples of stimuli that go beyond conditions to which we adapted, and thus can overload our natural reward circuitry.

Addictions can develop into predations – adding another level of maladaptation that involves more than just unwanted or out-of-control activations. Predations go further in self take-over to where a single motivational circuit usurps and practically kills one's self and all social sensibilities (respect for the self and rights of others, sense of humanity or morality, etc.). Excessive amounts of money, fame, power (economic, political, institutional, shared-attention, or physical), anonymity, and/or opportunity can render a person vulnerable to engaging in predation (either legal or illegal). Sexual predation sets in for some.

Example #3 – Grief and Attachment

Hunters had just killed the baby gorilla's mother. The baby lay there pounding his mother's lifeless body, screaming at her, moaning, and seeming to beg her to wake up. Finally, the baby's body went limp. His heartbeat went down and down, slowing way below normal. Another female guerilla picked him up and held him to her chest. He 'lit into her', beating his little hands into her, sobbing, screaming. She remained calm and peaceful, as if she understood what he was going through and knew he needed to process his loss/grief and anger. Gradually he calmed down. Then they gazed at each other, as if they were mutually granting each other a place in their heart (i.e., developing an attachment).

Humans have comparable trauma- & loss-processing and attachment needs, but considerably greater brain plasticity and adaptability, than our near cousins.

A gorilla baby will die if not adopted and all gorillas have less adaptability to migration and relocation than humans. A human baby may not die, but it has needs that must be met to minimize developmental and mental disorders. The human baby's needs can be met through adoption; attachment-, self-, and sensory-motor development remediation; and trauma-processing, while settling into a secure and healthy attachment, family, and habitat. Migration and relocation in humans is commonly accompanied by loss-, trauma-, and adaptation-processing needs.

Example #4 – Breathing

There is a place in your mind that is quiet, peaceful, and calm. It's always been there, even though you may never have noticed it and may rarely go there. You get there by simply watching your breath…

Sound familiar? A variation of this exercise is practiced as part of many meditative traditions, healing rituals, stress relief routines, prayers, and psychotherapies. So what has it got to do with the brain? In the first part of this chapter I explain that the brain grows not as mere neurons, abilities,

Last chapter diced and sliced the lower as functions, needs, and reaction circuits. The brain grows and functions as a *self*, its various parts integrated and attached – up and down and side to side. Watching the breath connects us to our embodied lower self. When we calm our breath in the here and now, it takes us out of the thinking and 'cortical processing overdrive' activated higher in the brain.

Many people are familiar with using body processes – including eating, breathing, exercise, posture, and facial and body gestures – to bring about improvements from the bottom up. Practicing preferred behaviors, habits, and routines are well-known contributors to improvement. Do you know, for example, that standing in a posture that feels bodily confident, actually cues the brain to feel mentally confident?

Example #5 – Excess/Poverty, Structuring Healthy Balanced Rhythms

In addition to distorted payoffs, the patterns in science seem to show that both 'under having' and 'over having' cause problems for the individual. The character Scrooge in Charles Dickens' novel is an illuminating illustration of a person whose asocial has drowned out his moral and humanistic sensibilities (in this case, Christian values – A Christmas Carol). He may even have an acquisition disorder – a compulsion to acquire money; an obsession with having, counting, and hoarding money. Opportunity and exposure to excessive supply payoffs has blotted out his love, joy, and care for not just his employees and their families, but also for his own family

Conversely, under-exposure to stuff required to meet your basic needs can lead to it's own disorders. Poverty can affect your ability to be selective in where and to what you pay attention, contributes to decreased brain activity in the prefrontal cortex, and may even hinder brain development – especially in children below age 10.

Even if parents recover, falling into homelessness has been shown to effect children under 10 in this way, even long after their family's recovery.

If you limit a young child' access to junk foods and present them with a variety of healthy foods, as well as model enjoying interesting food as an adult, they will naturally select the nutrients their body needs and not fall into excess or deprivation.

Example #6 – Play

The cat practices her escape route from the family dog. The three young monkeys replay the scene of their almost-capture by the lion over and over. They are doing the work of play, emotionally mastering the trauma they survived. Play is an essential component of social, emotional, and cognitive evolution and an essential process of mammalian learning. It is intrinsically enjoyable as well as healthy and adaptive. Adults would benefit from playing at singing, dancing, poetry, writing, and dramatic acting in real life play, not videogames and vicariously.

Threats to Survival

An important thing to remember about lower is that extreme experiences are activating the perception of threat to basic survival:

- ◆ Physical survival (death)
- ◆ Social survival (loss, abandonment, thrown away)
- ◆ Self survival (loss, annihilation, engulfment)

The Emotion of Caring

There is one notable exception in the above list. Mammals giving care (i.e., nurturance/protection/modeling/training) is an upper function, as I introduced in the last chapter, while mammals receiving care is a lower function. Let me explain.

I classify the giving of mammalian nurturance as an upper function. The mammalian circuit of giving nurturance/care is rooted early in evolution in the lower care-taking roles that many animals possess. However in most mammals, the caregiver relationship switches from instinct-only care to an emotional attachment system, which counts as upper. Caregivers provide nurturing; protection; and modeling, training, coaching, education, and other preparation for adult roles. A child builds its upper (via the lower-upper construction process) based on these outer experiences. Thus, receiving mammalian nurturance is lower.

It is the experience of lower receiving nurturance from outer that enables the growing individual to give nurturance to him or her self. Thus, humans build a right-sided self-in-attachment sense of self, which will be further described in Chapter Seven.

A 'feel good' benefit of the earlier in evolution roots remains: The enjoyment of giving as well as receiving nurturance. Caring/guiding/protecting, such as mother-child, with its release of the natural feel-good/pain-reliever hormone oxytocin, flows in both participants – the giver and receiver. Parent-child, pet-human, adult pair bonding, and sexual orgasm, likewise tap into this natural feel-good flow.

Options In How We Identify With, Experience, and Remedy Lower

Any given person has options to understand lower aspects of self-functioning. Examples of the options are below. Sometimes it is useful to understand them as parts of self – and the further option to make sense of them as more childlike or more as adult reactions, appetites, physical needs, and emotions. Sometimes it is important to understand that the whole self has unintentionally identified with, experienced being trapped or taken over.

Examples for Firefighters, Police, and Military in Stress Recovery

Firefighters, police, and military members do jobs that encounter death, wounding, and destruction – and the threat of such – in the course of duty.

A stress injury can occur much like the shrapnel wound from the explosive device. Army Chief of Staff Gen. George Casey commented that "combat is inherently brutal and difficult, and it impacts humans in different ways" (Army News Service, 2007). There is comparable brutality exposure in police work.

For some, the best help comes from framing the stress as a lower, almost inadvertent, jitteriness that a person and his/her loved ones mobilize together to calm, just as they would calm a baby that is just jittery. In this frame, "just weeks of dedicated, regular practice – especially if done some of the time with a caring family member – balances and starts to settle the biochemical swings and irritated nervous networks which fuel much of mild and some moderate PTSD."

Practice with: guided imagery, conscious breathing, progressive muscle relaxation, mindfulness meditation, yoga, qigong, self-acupressure, prayer. Other methods such as therapeutic massage, energy work, such as Reiki or Therapeutic Touch, aerobic exercise, listening to music, mindful walking in the woods, working with art or gardening may help.

Alternatively, the problem may best be addressed as an activation of the fear circuits stuck in trauma. It is the person's lower fear circuits, which were originally activated in specific habitat conditions. Some psychotherapy methods would this employ exposure in imagery or virtual reality to those conditions and cues, while the person is being present with the incompatible information and experience that they are in a safe place and time now.

Sometimes the injury is more serious. "Sometimes "the mental, physical, and emotional changes may reach a culmination point when an individual's internal resources and ability to sustain him [/her] self are exceeded by the demands and stresses" confronted, Gen. George Casey wrote.

Some experiences shift a person in the less desired direction and they haven't yet been able to shake it. The lousy has taken over too much of neural territory and the person needs himself or herself back. For example, one soldier wrote, "The more tours of duty I do, the more of Afghanistan there is in me. The less of home is in me," said one soldier in the "Tattooed Under Fire" documentary.

Other Adult Trauma Examples, Including from Accidents, Sports Injuries, and Medical Procedures

Examples of additional events adults have experienced as trauma: Participation in or witnessing catastrophic military actions (e.g., watching someone get blown up next to you), assault, robbery, rape, incest, other sexual abuse, physical abuse/beating, spousal abuse, accidents (vehicle, motorcycle, bicycle, falls, sports injuries, falling through a glass window or off a ledge), injuries, death or threatened death of a loved one, suicide of a close person, life-threatening illness, living through a hurricane or tsunami.

Many a professional sports career has come to an end when the athlete cannot shake a case of the *yips*. The self has not recovered the amazing capacity to adapt to a hard ball whizzing past the head at 90 mph, feeling no fear (a.k.a. normal dissociation).

It is not uncommon for a person to experience intrusive medical procedures, including surgeries, as a trauma. What was expected to be a routine dental or gynecological procedure causes, for some, intruding nightmares or flashbacks in the weeks afterward. The drawing of blood causes another to faint.

Perception of the threat/possibility of harm, death, rape, molestation, loss, kidnapping, or abandonment in a specific real situation or circumstance may trigger the same survival response. In other words, the alarm centers of the brain are assessing or thinking, "this could lead to me dying."

Unfortunately for humans, many of the psychological, social, or economic slings and arrows of life can trigger this same biological sequence. Health crises or loss of home, job, health insurance, significant friends, family, or community may trigger traumatic survival reactions in the brain.

Lower Self, Remedies For Lower

Now the amazing fact is, that our lower is not alone. Lower wants to know what upper you thinks; wants your guidance. It needs to know your ideas, beliefs, loves, loyalties, constructs, store of past experience, and anticipated and desired future. Some of this knowledge and communication from upper you to lower you happens internally and automatically. Regardless, most of us do not realize how happy it makes lower us feel when upper us talks to it/him/her. Lower us can feel so relieved, it is surprising. We can supplement every day with conscious tuning in to, describing, explaining, and requesting from our lower self.

Further, when lower is not alone, any trauma or developmental need not yet met, can be recovered from or remedied. Lower has upper you, your loved ones, friends, and community healers. Subsequent chapters will explore our universal best adult selves, our highest potentials and capacities, the range of upper and outer available to lower – and available to us to communicate, guide, attend to, and structure lower us. The last chapters illuminate our powers to draw upon and strengthen our self and readapt.

Five

Design Change:
Sense-Perception, Motor-Action, and Body In Upper

"The best road maps to human bodies lie in the bodies of other animals. ... The reason is that *the bodies of these creatures are often simpler versions of ours.*" In particular, "reptiles are a real help with the structure of the brain." (Shubin, 2008)

The Sense-Perception and Motor-Action Systems

The non-sided brainstem plus the basal ganglia constitute the human brain's structural base, which we have roughly in common with reptiles. One way to understand the major structural design changes from our reptile-like ancestors' brains to our own is to consider our sense-perception and motor-action capacities.

As you learned in Chapter Three, reptiles have all their sense-perception and motor-action powers right there on the same level as the rest of their brain and body. Evolution redesigned this reptilian single unit into separate multilevel mammalian physical-sensory and motor-action systems. Using the same house analogy as in Chapter Two, the new physical-sensory and motor-action apparati are like a section of that two story house without the floor between the stories, for greater connectivity and accessibility. Our sense-perception and motor-action systems start out lower, but require upper integration; we cannot walk (motor-actions) or see color (sense-perception) until we grow upper levels to these systems.

The mammalian upper includes partial placement of these sense-perception and motor-action functions. The change to upper enabled organisms to take in, process, and internally represent/code larger quantities of more complex information, which led to more complex experiences and greater involvement of the outer in upper construction. All this complex information and the many ways outer shaped upper development can then influence what we do (motor-actions) and what we sense-perceive.

Procedural Memories

Many automatic behaviors of the physical-sensory and motor-action systems involve some of what science calls our largely unconscious procedural memories. You and I know them as:

* Walking, driving, cycling, swimming
* Complex sports and creative arts routines/training
* Wake up and going-to-sleep routines
* The response procedure of a guard in an emergency situation
* Ways community police are trained
* Cooking, eating, cleaning, hygiene, sex routines

Have you ever been not quite awake in the morning and do a routine pouring motion only to look down and notice you had poured it into the wrong container? Coffee in the cereal bowl? Milk in the sugar container? That happens when your more-back of the cortex sense-perception and motor-action routines (procedural memories) began performing before your highest, most nuanced front of the cortex centers booted up to give you more fine-tuned guidance on where to pour what.

The Rest of Your Self and Social

Importantly, the person, family, and society all play roles in formulating elements of procedural memories, which become second nature.   For example, training to execute response procedures in a threat-to-life condition also largely influences our reactions. Military, police, EMT, and other first responders have such trainings as additional tools in their emergency response repertoire. Preparatory training can be of higher or lower quality and vary in what community values and social sensibilities are structured within the procedures. Urban police trained in learning about and connecting with a community of their own as part of a mission to keep the peace, have different experiences than urban police trained as if they are in a war zone to take over a town in foreign territory. Recently, the news presented a contrast in police training for when a gunman attempts to go on a killing rampage. The previous police training had officers wait outside while calling for backup. The new training has them immediately rush in, identify the perpetrator, and take him out.

Our intricate whole includes:

1. Emotional and motivational lower reactions,
2. Upper judgment, criticism, and evaluation reactions, and
3. Sensory-motor performance activity and procedural memories.

Each aspect can react/perform separately or together. This ability gives humans the option to bypass or isolate certain reactions – like a sports performance can be with or without emotion. Golf seems to be an ideal scenario in which the almost solo individual can practice getting into the flow of performance, leaving emotion and social observation out of it. Pre-prepared physical-sensory and motor-action system procedural memories, along with the roles and rules of the team game, allow the baseball player to come within inches of a 90 mph speeding projectile and not react with "the yips". The new military recruit fires crazed with fear at his/her first exposure to enemy fire; the experienced soldier calmly relies on and follows the trained procedures and rules of engagement.

Are they you?

Humans may consider procedural memories to be part of their identity or not. If you like your reactions and procedural memories, use them to enhance your identity. As a self, you have flexibility and options in how personally you experience these skills. They may be just skills, be part of your identity, bother you, or bring you joy and payoffs.

Taking a self-defense class can be considered learning procedures and skills – part of your talent identity. Alternatively, some procedural training experiences can add environmental options to your self identity, such being able to assume an 'alertness and defend' state. You have been taught to assert and defend yourself when necessary – does that become part of your prevailing identity?

How we experience these sense-perception and motor-action components also depends on how they are coded and patterned in our families, social communities, and other habitat context, which is discussed periodically throughout the remainder of this book.

Body Is Up There Too

In Chapter Three, I described how the mouse has an upper self-constructed map based on the mouse's individual experiences, which includes representations of its body in the hippocampus. The mouse experiences particular aspects of its habitat in association with specific parts and functions of its body. These habitat-body associations neurologically connect in the mouse's hippocampus.

Similarly, multiple places in the human brain – such as the hippocampus and the sensory and motor cortices – routinely form upper representations (also called maps, templates, or grids). In short, we map experiences of habitat in association with representations of involved body aspects.

Now I want to add one more element to my description of the major structural design change concerning our sensing and motor action capacities at the beginning of this chapter. We have a somatic sensory component to body physicals in the cortex. For example we have mapped representations of the complete body (called somatotopic maps) to keep track of and process various components of touch in the cortex. We have other sensory processing areas, such as those to attend to and process balance, sense of body space/place, pain, and the chemical senses of taste and smell. In fact, the whole motor and sensory systems of the brain are arranged somatotopically, with specifics regions of the cortex being responsible for different areas of the body, such as depicted in this graphic:

Image 4

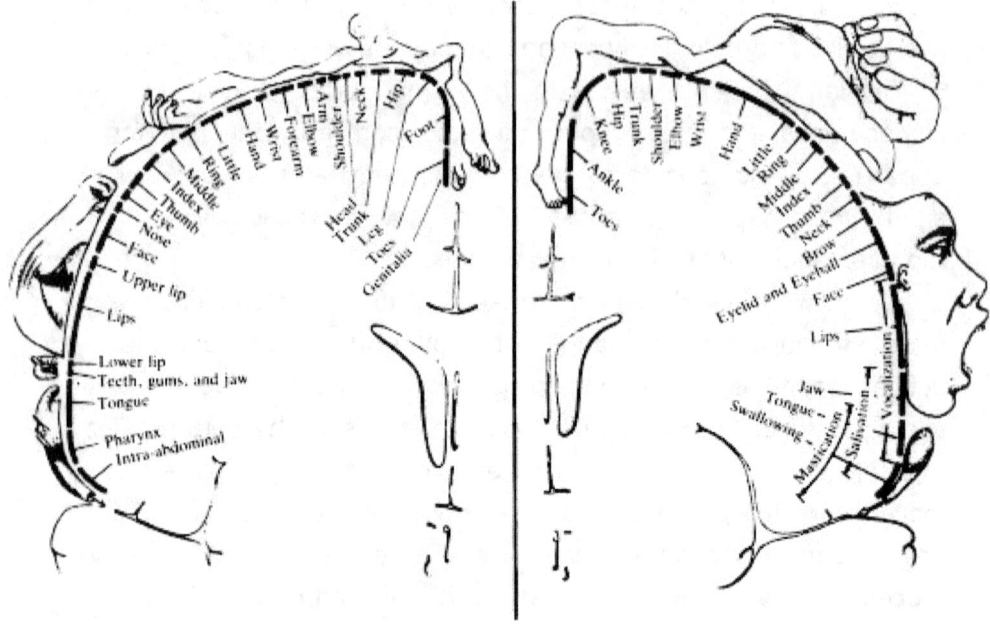

Phenomena such as phantom limb pain occur because the body has both lower and upper places in our self system. Once a limb has been surgically removed, the brain has to take that lower information and change its upper representation, otherwise upper still experiences the leg as there and in pain.

The brain also reads aspects of the body in daily routine. Your own body action, gestures, posture, stances, sensory-motor performance, motion, exercise, smiles/frowns, physical problems or health (along with food nutrients consumed) are important lower components of self presence and well being. The brain uses body information, as well as upper representations, when formulating its mood and planned activities/urges.

To illustrate this relationship, how you stand – straight up, martial arts position, yoga pose, slouched, or whatnot – can determine your mood. Raised arms and upper body workouts give a message to the brain that the body is involved in vigorous work. This makes the brain feel good and support you being in a good mood to do the work, partially because the brain anticipates that doing such work is associated with coming rewards (as was true in our Environment of Evolutionary Adaptedness, or EEA; see Chapter Six). Go ahead: raise your arms now. Simple body actions can sometimes improve your mood. Moving a body part does not just affect your position or muscle action, it also changes what signals your body sends to your brain – and therefore what upper does with that information, in terms of the whole's current agenda and well-being. In the human EEA, if our bodies were moving or our arms uplifted, we were probably engaged in some activities from which we and the group stood to benefit, and this movement made the brain feel better. Sports and game winners frequently raise their arms in a triumphant gesture, to give one modern example of this phenomenon.

Sometimes people use food to help themselves feel better – especially rich, sweet, fatty, high-nutrient, or 'whole' foods. Your brain is designed to think that consuming these foods is always positive. As it evolved in the EEA where such foods were scarce and highly desirable, your brain automatically releases reward 'feel good' hormones when you eat them. This demonstrates how outer (food) can affect lower (body), and then upper reads lower's experience (getting fed). In this case, upper then releases 'feel good' hormones and otherwise works to improve your mood. Some people have even reported that eating certain foods can calm migraines.

Before we go on to the next chapter, the reader might be interested in the fact that this overall multilevel design change contributes to the evolution of our cortical space, which is the brain platform for our abilities to imagine, mentally rehearse, fantasize, pre-fashion, and refashion ideas, behaviors, and scenarios. These advanced capacities emerged in associational areas of our cortex – 'mixing grounds' of sensory perception, body maps, and motor action pre-activation that are more in the middle of our cortex.

Related abilities that also emerged include that we can imagine/plan to activate, or with mirror neurons, can activate practicing or as-if we're doing an action we watch another doing or as-if we are in the drama/circumstance the other is in. Such observation activates all the same neural networks as if your brain and body are doing something or being in a circumstance, without actually having to do the action or be there. Psychotherapists draw upon this capacity for many important reasons: empathy, to get an understanding of what it is like to be in the other's circumstances, while not overly identifying or being vicariously traumatized by another's trauma; and to be able to internally associate to possibilities for needed perspective, action, or healing process participation for the client.

Six

Upper, Complexity

Imagine hearing this lullaby sung in calm and soothing tones:

Lullaby and good night, with roses bedight [delight];
With lilies o'er spread is baby's wee bed.
Lay thee down now and rest, may thy slumber be blessed.
Lay thee down now and rest, may thy slumber be blessed.

Parents singing their children lullabies demonstrates how outer can have a calming, perhaps sleep-aiding, effect on lower's construction of upper. Outer experiences essentially become upper, which has differentiated right and left sides. In the case of a lullaby, the left understands the words, while the right processes tones and learns to reproduce the calming effect without outside stimuli.

Later in life, the ability to regulate and soothe ourselves is possible because we first experienced caregivers calming us from the outside. Eventually, we became able to soothe ourselves with internalized copies of significant others doing it for us.

But remember upper is a quasi-location generalization for all of our cognitive capacities, for putting our 'thinking cap on', for an extremely complex array of capacities. All model constructions of habitat, ecosystems, and the world that allows animals (including humans) to negotiate their environments adaptively are upper guiding lower functions. Beliefs, conclusions, and models of self, other, self-in-attachment, social-groups, and people-in-general are upper constructions. We humans have an extremely complex array of self adaptations, models, constructs, and instructions guiding what we do.

Our brain expects that our higher functions need to be integrated into our existing biological being, such as our physical (lower asocial) rhythms/survival and emotions – including play. Such integration allows for processing emotional/social reactions and exercising complex mental talents and interests – e.g. imagination, story-telling, understanding cause/effect, exploration, scientific experiments, and sensory-motor abilities like throwing a ball. (Note: As an aside, in modern society, mastering primary cognitive, social, and emotional skills and sensibilities for adult life involves educational systems and related 'extra-curricular' activities.)

Recall the Chapter Two list of 'basic' upper functions and capacities (**Table 1**, partially represented here). Psychologists refer to these with concepts such as cognitive, social, and regulating/managing of lower's motivations and emotions. In initial lower-upper construction, lower grew a cortical that includes lower's experiences of outer. Some of this construction participates in any needed adult adaptation, rehabilitation, or other improvement.

Table 5: Upper Functions and Capacities

<u>Upper</u>
Mostly the Cortex

♦ Thinking, Cognitive Function, Ideas, Imagination, Language
♦ Representational Ability, Symbolic Capacity, Theory of Mind
♦ Beliefs, Moral Ideas/Values, Humanistic and Social/Caring Sensibilities
♦ Planning, Reason, Will, Conscious Decision-Making
♦ "We" Intentionality
♦ Concepts, Frameworks, Constructs, Theories making sense of inner & outer information, events, experiences, narratives
♦ Causal Belief, Interpreter
♦ Effective Procedures for Performance, Success
♦ Mirror Neurons (most)
♦ Conscious Awareness/Attention, Observing (all outer & inner)
♦ Sense of Selfhood
♦ Self Awareness, Mindfulness, Mindsight, Meditation
♦ Copies of Body Components
♦ Upper Memories (conscious and/or unconscious); storage space
♦ Sense-Perception & Motor-Action Integration Functions
♦ Emotion Integration Functions
♦ Motivation (instincts and urges) Integration Functions
♦ Self Management & Control (scientists' term: self-regulation)
♦ Observational and Vicarious Learning
♦ Learning by Teaching
♦ Increased Habitat Flexibility and Adaptability
♦ Final Attention and Performance Self Stage Managers (cerebellum/new brainstem)

Lower needs its upper for such capacities and sensibilities as self-control, cognitive skills, guidance, planning, information about present/past/anticipated future, regulation, and orienting beliefs or ideas about self/attachment/family/ community/culture. Vice versa, upper receives information about lower's needs, states, and reactions. 'Conscious us' gets the signal that we are hungry or need a bathroom break – or not.

We can understand the effect of upper processes on lower through a comparison of gorillas and humans processing emotions such as anger and grief. On the one hand, gorillas have a simpler cortical structure, which provides them with less cognitive abilities and less options and flexibility of responses. On the other hand, they more naturally and effortlessly participate in the biological processes for mental health recovery available to them.

In humans, upper can interact with lower in different ways. One option entails the child's brain-mind-self participating in processing/healing from loss/trauma, like a baby gorilla. Another option involves the child suppressing its feelings. While humans' cognitive capacities enable us to more-easily survive without fully processing traumatic events, not processing traumas and losses results in more vulnerability to chronic mental and physical health problems as well as constricted functioning (i.e. avoiding certain scenarios/behaviors). These dynamics are further discussed in the attachment section of the next chapter.

The following example illustrates the roles of upper and outer during the adolescent transition:

One of my clients, a twelve-year old boy, found that all-of-a-sudden he was extremely mad at his mother – yelling and angry at her constantly. The son was not nearly as mad at his father, so he concluded that his mother must be doing something wrong to make him feel so mad at her – and told her so. In our next session, we discussed an alternative possibility: that his reaction is a normal part of forming an independent identity. I told him about the process of pfc neuron pruning (dropping away), which encourages teens to incorporate what works from their experience in formulating a sense of themselves as both separate from and connected with their parents.

Because he seemed to have had been closer to his mother than his father before, I offered the hypothesis to both him and his mother that it was harder for him to establish independence from someone with which he has thus far been more dependent. Thus, without understanding the biology involved, his brain found anger to be a helpful tool for driving the separation from his mother and establishing his own personal identity. This explanation rang true and the boy seemed almost relieved. The full brain-mind-self understanding of human nature enabled me to map the experience of this family to pinpoint what upper perspective information might be helpful for this child to consciously know. Internally, the son's brain-mind-self assessed the new information and determined it was relevant to understanding his emotions and perspective. After this realization, incidences of mother-son anger decreased significantly in that household. Note, all three parties – son, mother, and therapist – had no sense of any other self-information-meaning to his anger at his mother.

Humans' Complex Brain and Self

Biology classifies brains based on their complexity, and the human brain counts as not just complex, but "extremely complex" (Lewis, 2009). The expansion of the cortex via the lower-upper construction method enabled brains to evolve that extensive complexity.

In particular, the way that lower and earlier capacities – shaped by outer experiences – both grow and integrate with upper functioning includes important aspects unique to us. To that end, I present some preparatory orienting below.

Our Right and Left Brains Grow Up

Brain-mind-self child development has a specific right-then-left pattern of structured growth relating to both the asocial-social-asocial and lower-upper themes previously discussed. The initial asocial component is basically non-sided; subsequent social integration occurs more up the right side; and the higher asocial specializes upper left and/or in the prefrontal cortices (pfc) integration of both sides. Additionally, while organized bilaterally across vertebrates, upper functioning is more extremely lateralized in humans than any other primate. Right-sided self (the focus of the next chapter) grows from its experiences relating one-to-one (dyadic) and forming attachments within families and groups.

Adults function with an intrinsic left and right cooperation of our hemispheres, whose interwoven roles and specializations are choreographed by and through our pfc. This book intersperses references to our sided functioning – for example in my writing, I am participating in 'left hemisphere' analyzing and presenting of concepts and frames and 'right sided' focus on your experience of being your self. Within the latter focus, depending on what is useful in a particular situation, you can take any of three roles (previously discussed in Chapter One) in regards to your brain: driver, car, and co-mechanic. As a co-mechanic, you also may be experiencing your self from a more right-sided (you are doing the work) or left-sided (you are watching or analyzing the work being done) perspective.

The Pleistocene Hypothesis

The historical record thus far constructed with scientific inquiry required a change in classification recently. The term hominin now refers to modern humans and their bipedal ancestors (previously known as 'hominid', a category now known to include chimpanzees, bonobos, gorillas, and orangutans).   The hominin existence on earth begins in the Pleistocene Era (the most recent period of repeated glaciation from 2.5 million to about 12,000 BP*).   Around the same time as the Pleistocene Era (and hence the last Ice Age) began, the genus *Homo* diverged from other hominin species.   The *Homo* line correlates with the word "human", and the only extant (living) species in this genus is *Homo sapiens* – us – modern humans.

The speciation phase is when the unique biological characteristics of a species become fully established.   Humans achieved the genetic basis of our species during the Pleistocene Era, about 200,000 BP (when the scientific record first documents the appearance of anatomically modern humans – i.e. *Homo sapiens*).   Unique body changes likewise contributed to humanness, such as modifications of the mouth, larynx (voice box), and foot structure, which were present at speciation but likely continued to develop afterwards (such as through utilization changes).   *Homo sapiens* achieved behavioral modernity around 50,000 BP, meaning we had many capabilities at speciation that were not realized until later.

Some paleo-anthropologists and archeologists describe the rapid period that led to behavioral modernity as a "Great Leap Forward."   By this point, our ancestors roughly possessed the biological and genetic capacities for: complex grammar language, symbol utilization, internal representations, theory of mind, causal belief, learning by teaching, 'we' intentionality, sense of a separate self with interdependence, increased habitat adaptability, and greatly expanded adult neuroplasticity.

---

* BP (Before Present) is a time scale many scientific disciplines employ to specify when events in the past occurred. Standard practice dates January 1st, 1950 (AD or CE) as the arbitrary origin of the BP scale.

Additionally, during this relatively short period of time, our ancestors rapidly learned to produce artifacts beyond tools (e.g. make animal hides into clothes), bury the dead, paint pictures on walls, and travel/migrate to different environments. Language was certainly advanced by this period, though it developed earlier and was drastically shaped by interpersonal interactions, collective formulating/learning, and participation in shared group practices interacting with material objects – i.e. culture. That interaction of biological capacity and cultural practice is the subject of cross-disciplinary studies including psychology, neuroscience, archeology, paleo-anthropology, and evolutionary biology.

While existing evidence shows slightly different origin dates for some particular traits/behaviors and continued research may specify further differences, an overall pattern remains observable in the record.

Around the end of the Pleistocene Era – which corresponds not only to the end of the last Ice Age, but also the end of the archeological Paleolithic Age – our biological capacities, social organization, interaction with each other, and individual/collective interaction with material objects, together shaped the accomplishment of a turning point in human evolution: the ability to control the environment through agriculture and thus establish permanent settlements. Estimates as to when these developments occurred tend to be between 10,000 and 13,000 BP, and they distinguish what archeologists and anthropologists refer to as the culture of modern humans. A term not referencing current modern society, 'the culture of modern humans' started appearing in the archeological record in prehistory (defined as before written history), and is distinct from previous human culture for the above listed reasons.

Look at the following Table. Our interpersonal and social-organization processes (certainly after Homo erectus achieved pair bonding) remained stable throughout the epochs of speciation until the dawn of 'modern human culture'. Remember, EEA is *Homo sapiens'* environment of evolutionary adaptation.

Table 6: Stages of Human Evolution and Social Organization

Note: All stages share stable Social Organization & Interpersonal Processes, except the last one.
⇓

Approx. Timeline	Stage in Human Evolution	Social Organization & Interpersonal Processes	Notable Signature Human Qualities
2.5 million BP	Genus *Homo* diverges from other hominins	• 1-to-1 personal attachment relationships among children and parents. • EEA form of interpersonal interaction/communication, work & resource distribution, interaction with material objects, and leadership/ followership. • Collective formulating-models/learning and participation in shared group practices interacting with material objects	• Larger brains *Homo erectus achieved:* • *Cooking with fire* • *Pair bonding*

Approx. Timeline	Stage in Human Evolution	Social Organization & Interpersonal Processes	Notable Signature Human Qualities
200,000 BP	*Homo sapiens'* speciation completed	• 1-to-1 personal attachment relationships among children and parents. • EEA form of interpersonal interaction/communication, work & resource distribution, interaction with material objects, and leadership/ followership. • Collective formulating-models/learning and participation in shared group practices interacting with material objects	• Anatomically-modern human appearance • At birth: human genome and basic brain hardware • Biological capacity for behavioral modernity

Approx. Timeline	Stage in Human Evolution	Social Organization & Interpersonal Processes	Notable Signature Human Qualities
40,000-60,000 BP	Behavioral modernity achieved – 'great leap forward'	• 1-to-1 personal attachment relationships among children and parents. • EEA form of interpersonal interaction/communication, work & resource distribution, interaction with material objects, and leadership/followership. • Collective formulating-models/learning and participation in shared group practices interacting with material objects	• Complex grammar language • Symbol utilization • Internal representations • Theory of mind • Causal belief • Learning by teaching • 'We' intentionality • Sense of a separate self, with interdependence • Increased habitat adaptability • Greatly expanded adult neuroplasticity • Produce artifacts beyond tools (e.g. make animal hides into clothes) • Bury the dead • Paint pictures on walls • Travel/migrate to different habitats
10,000-13,000 BP	'Culture of modern humans' begins	• Social organization fundamentally disrupted	• Domestication of plants and animals • Large settled societies • Radically different relationship with habitat

Throughout human evolution before the 'culture of modern humans," human interpersonal and social-organization processes included basics, here described in two categories:

1. One-to-one personal attachment relationships, especially among children and parents, and
2. EEA form of interpersonal interaction/communication, work & resource distribution, interaction with material objects, and leadership/ followership. Collective formulating-models/learning and participation in shared group practices interacting with material objects

It seems that important brain-based, social- and developmental-psychological basics were stably present during that whole long EEA epoch. To understand this point, combine:

1. Brain-mind-self framework understanding of key broad patterns in the evolution and development of the human brain with
2. These social-organization/interpersonal-processes present not just at speciation, but stably present until its breakdown with the domestication of plants/animals and living in large societies.

Together they point to an intriguing hypothesis. Because our EEA and these interpersonal and social-organization processes were extant throughout the Pleistocene Era, I label this as: ⇒

The Pleistocene Hypothesis

Our very unique EEA forms of interpersonal and social-organization processes provided a stable base for our latest brain evolution and development and the necessary social conditions that enabled:

* The very evolution of Homo sapiens,
* Our latest-in-evolution qualities, including our flexibility, adaptability, and greatly increased adult neuroplasticity,
* Sapiens' 'great leap forward'
* Our habitat-changing accomplishments (including agriculture and mass-settlements)

This latter human transition happened at the end of the Pleistocene Era and started the 'culture of modern humans', wherein our relationship with our habitat (outer) fundamentally changed. Ever since, human societies have become markedly different, and our universal biological functioning and capacities no longer work so consistently in our favor.

Of the two interpersonal and social-organization processes mentioned above, Chapter Seven will detail the first and Chapter Eight contains the second. All upper, the subsequent six chapters add: two modes of thinking we employ in these experiences, a description of our self adaptive functioning, our more left hemisphere and latest-in-evolution cognitive and intellectual self powers, the disruptions in social organization self experiences in the 'culture of modern humans', our prevailing self functioning (whole), and the tremendous neuroplastic/adaptive recovery and repair powers we have.

Seven

The Self Grows Up – Right Hemisphere Self

Consider this scene:

A cannon fires at a war reenactment battle. The sleeping baby in the arms of its mother initially startles, opens her eyes, stares into her mother's face, then settles back to sleep, unconcerned about the cannon firing around her.

When a cannon fires, a dog's brain will react by becoming upset and unable to be comforted, no matter how good a relationship it has with its parent/owner. However, because a human baby can become securely attached to its caregivers, when the baby sees that the caregiver is *not* distressed by the cannon fire, the human baby settles down, assured that the loud noise is not a threat. Likewise, when a baby does become distressed and other needs have been taken care of, a caregiver can comfort the little one to convey that things are now alright, such as through rocking or singing a lullaby (like the example at the beginning of the previous chapter).

Attachment – a key element of early postnatal self and brain development – has an important role in the upper integration of the early lower brain. A human baby's self-functioning development will vary depending on the presence and physical contact of the caregiver (to whom they are securely-attached) as well as how that figure responds to a noise or other potential threat and takes care of both him/herself and the child.

Impersonally stated, when the lower constructs the upper right brain, it forms attachments (a felt/body sense of personal connection/love/care) with people components of outer that it copies in upper. These copies embody: external individuals/groups, the individual's felt/bodily sense of relating to them, and any associated automatic conclusions/beliefs about self, others, and the world (such as in terms of goodness, mattering, being safe, or being trustworthy).

You, your brain, your primary attachment(s), and relevant outer all contribute to postnatally constructing you, which first grows more up the right side.

In Chapter Two, you learned about right cortex functions and capacities that were more cognitive. The right cortex's role for you consists of more than just the above skills (and all its sided-shared cortical capacities) added to lower emotions and motivations – right-sided development initiates a person forging a sense of selfhood, a.k.a. a brain-based self structure. A newborn is a whole that knows who s/he is and who s/he can be – and finds ways to develop as best s/he can through to an adult. In the process, his/her lower becomes much more interwired up the brain's right side than the left side, hence the particular importance of initial non-verbal right hemisphere development.

Secure Infant Self-in-Attachment Development

Attachment researchers would probably use wording like 'infant self development in a securely attached relationship'. My phrase emphasizes the self's development enabled by her/his relationship. This initial more right-sided self-organizing of the brain occurs through young children forming attachment relationships with other people – other brains, minds, and bodies – with whom they have attuned and positive relationships (for example, Schore, 2009). Secure attachment happens one-to-one (dyadic), as a fully-present parent/caregiver tunes in, pays attention, and takes care of things from the baby's point of view (attunement), while independently meeting her/his own needs (like rest and adult contact). Attachments involve communication between the infant's right-brain (lower growing to include upper) and the caregiver's right-brain (lower and upper), as well as making internal copies of outer.

It is amazing how much a 4-6 month old can communicate non-verbally – can read the non-verbal cues of mother and other significant adults and attempt to communicate with them using body or mouth gestures. One new child-rearing technique even teaches human infants sign language to prevent unnecessary mis-attunement between caregiver and infant.

The very first thing a newborn needs to do is establish its primary attachment(s). Children use attachments to grow a right-sided sense of self – a self structure. Through secure attachments, the baby discovers safe and viable life rhythms that support her/his growth and development. Mental health practitioners and developmental researchers sometimes call this a self-in-attachment structure because a person's sense of selfhood is built with early and continued attachments. (Thanks to our front-of-the-front cortex, the pfc, disrupted or disorders selfhood construction can be rebuilt, or otherwise repaired in adulthood. Like many basic human capacities, however, it is easiest to master in the early years.)

Other chapters highlight the importance of the environment of the parents. Here we just note that, from a biological point of view, the infant needs the parents/caregivers to secure supplies and assistance for themselves as well as the baby. This security ensures that the baby can regulate and recover from any feeling reactions or bodily distress fired in his/her existing circuitry in response to a threat. The infant also needs the parent to regain and resume safe routines for satisfying needs, if they are disrupted.

Note: It is not unusual for a person to find it hard to read the sections on attachment – or some other section – as just reading words can trigger old parts of your self to react... you may find yourself falling asleep or having more trouble absorbing information. When I first was learning about attachment, I realized that I kept falling asleep or blanking out at conference presentation on certain attachment difficulties similar to mine. Luckily, an adult can earn secure attachment status in psychotherapy and I did so. Whether initially accomplished or earned in adult development, an organized secure internal attachment status and self-other rapprochement capacity (ability to both be there for self and other).

Later Development

A second parent, sibling, extended family member, nanny, childcare person, neighbor, or other significant personal people in the child's life all form their own one-to-one attachment with the child. Human parents, nannies, grandparents, parent surrogates, and allomothers in human societies actually may be physically present for the child – and thus performing an important attachment role – even while key elements for upper, integrated, then individuated self-development are not happening.

Later development adds more layers of experience and new capacities as the individual self incorporates beneficial encounters and remodels inner glitches stemming from unhealed lousy experiences, traumas, or mal-adaptations.

The Self's Creative Adaptations to Attachment and Self
Development Difficulties

Attachment is too important for building the structure of the brain
and the self to do without.

Consider the Gorilla example previously mentioned in Chapter
Four: Hunters had just killed the baby gorilla's mother. The baby
lay there pounding his mother's lifeless body, screaming at her,
moaning, and seeming to beg her to wake up. Finally, the baby's
body went limp. His heartbeat went down and down, slowing way
below normal. Another female gorilla picked him up and held him to
her chest. He opened up on her – beating his little hands into her,
sobbing, screaming. She remained calm and peaceful, as if she
understood what he was going through. Gradually he calmed down
and she looked into his eyes – they began to form an attachment.

Gorillas do not have as much going on as humans. Their social
and habitat environment is simpler. In the cortex, as mentioned in
the last chapter, their sides are not as differentiated, nor do they
have our greatly expanded pfc.

On the one hand, their simpler structure provides them less
options and flexibility; on the other hand, they more naturally and
effortlessly participate in the biological processes for mental health
recovery available to them.

The baby gorilla will die, if another mother does not soon adopt
it.   Both new gorilla mother and baby intrinsically participate in
supporting the baby processing through sadness, grief, anger, and
other reactions he had to the loss of his original mother. The baby
gorilla's ability to bond with his new mother (form a new attachment)
stems from successful loss processing – which for humans can be
facilitated in psychotherapy.

Unlike gorillas that have very limited options to form
attachments and survive, a baby human probably will not die
without a mother, for it has much more flexibility in how to be

attached.   In addition to repairing damage from attachment disruptions and mis-attunements, human beings have the ability to create coping defenses thanks to our unique pfc. While humans can more-easily survive without fully processing traumatic events, such scenarios have a greater likelihood of causing chronic mental and physical health problems.

The effects of our capacity to cope with undesirable situations are both positive and negative, and can be seen throughout the variety of human existence.

Traumatic events happen, even if our social groups and other outer components try to minimize them. You know the old phrase: "What doesn't kill you makes you stronger"? Well, that's not always true. You will get 'stronger' only if you successfully process the trauma and fully repair any damage to your self. Otherwise, part of you may be stuck reacting to the trauma long after the triggering event is over, as part of yourself created an adaptation that made it easier to get through the trauma at the time, but afterwards hinders you in unexpected (sometimes unrealized) ways. Such adaptations help ensure your survival, but they also reduce your ability to fully live.

As already mentioned in Chapter Four, poverty and homelessness often affect children (especially under ten years of age) long after they are back in secure economic situations. During the extreme period, parents themselves may have an insecure or scared/panic state, such as worrying about securing basic necessities. The parent can be handling this with drinking, fighting, anger, overwork, as well as other mental health symptoms. Rates of abuse (spousal; physical, sexual or verbal abuse of children) and violence (suicide, homicide, criminal property damage/theft) go up. Children (especially younger) may learn this state and also develop insecure attachments with parents/caregivers who are not fully present for and meeting the needs of the child. Such a situation counts as a trauma that affects the child's development.

If close adults are insecure/upset/depressed and the child picks up on that, s/he may start feeling such reactions as well. The security of the parent may affect their ability to provide security for their child(ren).

A child who becomes preoccupied with the well-being of their parents/family may not be fully available for other areas like academics and extracurriculars. Conversely, insecurity – or worse – at home may drive the child to get away and thus focus extremely on school or other activities, even to the extent of avoiding or sabotaging relationships.

Consider the non-verbal attunement and acceptance the child needs. What does a depressed mother seem like to a young child? Not responsive? A mother constantly sick or with a headache, who is shut off from the child in another room. A parent who rejects a child or allows the child to be abused or hurt?

Practically speaking we come to have upper representations of the particular ways our lower self experienced how we were treated and how those caring for us handled us regarding the event (if the event was not at their hands), labeled by science with such words as resources, wise mind, and introjects.

Science can even predict the attachment style each parent will have with their not-yet-born child can be predicted ahead of time by assessing the parent's own current attachment style regarding their own early childhood parents (as the parents-to-be are experiencing it currently).   But do remember that attachment status security/organization and self structure may be disrupted or improved in the course of childhood and adulthood.

Right-sided self structure is about more than secure and organized attachment.   Right-sided development (in attachment, family, and safe habitat) in those first years establishes our capacities to regulate our lower emotions and instinctual motivations (here conceptualized as lower); to shift, manage, cope, and organize states; and to integrate and experience an internal sense of continuity/wholeness within our self. These – the building blocks of the self-resiliency – contribute to our later-in-life capacity to maintain/recover health and well-being when dealing with stress, trauma, and challenging circumstances. A person's basic sense of identity and selfhood– along with their beliefs, concepts, and conclusions about their self, place in other's lives, and safety and security in the world – start forming in these early life transactions.

Having a good early self development works to the advantage of your adult brain and body in several ways:

* First, the brain gets its own built-in attuned adult self that nurtures, protects, guides, and supports a person's self development.
* Second, this adult self is the foundation for essential capacities, such as those used in high performance and accomplishment situations and in sustaining/recovering health throughout the lifespan. Specific capacities include: emotion- and body- regulation, stress-coping, adaptability to rapidly changing environments, and healthy interpersonal participation.
* Third, these self-functions can be drawn upon, strengthened, and repaired to accomplish many coaching or psychotherapy goals.
* Fourth, an organized, secure, internal attachment status and self-other rapprochement capacity (ability to both be there for self and others).

Developmental Trauma

Now if you experienced the threat of death, loss, abandonment, annihilation or takeover of your budding self as you were developing, your clever brain-mind-self uses its lower construction and self adaptation powers to neurally split, dis-aggregate, or otherwise break apart your right sided self-in-attachment neural networks. Lower reactions such as urges to flee, hide, avoid, escape, fight, cocoon can be configured by the self with a copy of the lousy aspects of attachment figures, while urges to be close and nurtured become configured with copies of parents providing protection and care. Various coping reaction strategies become part of the memory networks. The self can use and abuse body functions and performances in coping with such trauma.

But, note that powerful mind, brain, and psychotherapy methods are available to optimize mental and physical health and body/performance realization. In addition to utilizing the relationship with the therapist (due to our profound self-in-attachment/social and development nature), it is our highest right and left and integrated powers and capacities that are drawn upon for healing.

Right Hemisphere Self/Intelligence

On your self-structure-based side you are becoming some awesome right hemisphere capacities. The self structuring you have accomplished intermingles with all your right cortex powers as were listed as **Table 3** in Chapter Two (with that half repeated here).

Table 7: Common Upper/Cortex Right-Sided
Psychological Processes

<u>Right Cortex</u>

•**Deductive Reasoning/Logic**
•Imagination, Visualization Images,
Fantasy, Artistic Ability
•**Non-Verbal Language** (tone,
accent, context, pragmatic) &
**Rudimentary Language
Processing**
•Facial Perception
•Spatial Manipulation
•**Approximate Number
Computation and Estimation**
•Sense-Perception and Motor-
Action Integration Functions
•**Extremes Of Emotion
Processing** (both positive and
negative)
•Intelligence for Emergency,
Chaos, and Novel Situations
•Innate Sense of Time and Space
Orientation
•**Sense of Selfhood**
•Experiential Associative Thinking
and Knowing
•**Conscious Awareness
Integrated with Body and
Emotions**
•Sensory Processing/Mixing
•Highest Level Of Wisdom and
Guidance, in Touch with What Can
Be Experienced as a Higher Power
•**Personal Felt Memories**
•**Long Term Memories**

All the capacities in the above chart can be instantly mobilized for you. There is anecdotal evidence that some persons who are stressed in preschool years, developed stronger right-sided intuitive intelligence.

A less lofty example is as follows: After a day of left hemisphere carefully cataloguing of facts and patterns, your big picture thinking and deductive logic right hemisphere is processing what you were working on figuring out. As you awake it can flood your brain with the answers and perspective that eluded you the previous night. Important spirit or higher power guidance for hard times is available.

Our left hemisphere goal-setter self can puts in a request to the right to use its powers to enable the whole to accomplish the goal. Our right truly wants to know what you'd like it to do for you and does so naturally, effortlessly and without complaint. It likes to do it.

Memories deemed by the self to be relevant to survival and thriving now are kept stored 'activation ready' on the right side and subcortical, ready to be instantly and associatively activated by experience. Memories the brain processes and realizes are really and truly over, it puts away as left-side fact memory.

Our highest self-development spirit and capacities are right sided. Look at the accomplishments of the two people in the Introduction of this book. Further, we each have the potential to connect to a deep and highest wisdom, goodness, guidance, and love, even in our darkest hours – however understood. For those who are spiritual or religious, this is an important part of self that connects us to God, Allah, Yahweh, The Good, Love, Christ, Peace, The Holy Spirit, Abbah, the collective unconscious, and whatever is good in the universe.

Eight

More About Social

The team in Afghanistan was in harsh conditions. It was hot, hard, dangerous, and much of the time very boring. Yet most reenlisted. They wanted to go back. Why? The on-the-ground reporter began to understand it was the security of the group bond. The soldiers were embedded in a community where everyone mutually cared about the members and goals; they were known, personal, valued; their individual work and skills were needed and expected; and they were given support to develop and improve – for the benefit of themselves and the whole.

<<0>>

The 6 month old was howling in the subway disturbing the passersby as the annoyed and overwhelmed 16-year-old mother got more and more irritated while she worked to change the baby's diaper. An older woman whispered into the mother's ear – they spoke back and forth, and shared a smile. Then the mother focused on her child and said: "It's okay sweetie. Mommy is just changing your diaper." The baby instantly stopped crying and the two began to enjoy a warm gaze. "I'm going to make you feel better. Your bottom will feel better when it's all clean and dry." A powerful and happy spirit was now beaming from that baby's eyes.

These two examples both demonstrate the importance of the social group.   In humans and higher primates, it is not just attachment and belonging to individual parents/families that young need. They – as well as their parents and families – are biologically wired to be in bonded social groups. Such tribes, bands, clans, collectives, and other associations are likewise represented/coded/memorized in the upper structures of the brain.

As mentioned in Chapter Seven, both adults and children need to be seen, acknowledged, and talked to as if we understand basic communication and are equivalent representations of humanity. Having others interact with us for the potential we are, is as essential for a person's spirit and development as food for the body. Science is piecing together that the brain has built-in assumptions for our entire social group about outer capably meeting such needs.

While the last chapter covered needs of the brain in terms of one-on-one or dyadic personal relating, this chapter emphasizes what science tells us about the brain-mind-self's needs and expectations regarding the social group.

Adults in Social Groups

Scientific findings in multiple disciplines are converging on an understanding of the social group characteristics that sustained humans in the stage often called the **environment of evolutionary adaptedness** (EEA). For *Homo sapiens*, the EEA ranges from the time the *Homo* line diverged from other hominins (as far as 2.5 million years ago) to before we began sustainably living in what archeologists and anthropologists refer to as the culture of modern humans (starting 10,000-13,000 BP). Our unique *Homo sapiens* speciation occurred during our EEA about 200,000 BP with the rise of anatomically modern humans. Our EEA involves habitat and culture universally shared by the ancestors of all living human beings today, before the Out of Africa dispersal 10,000-20,000 BP.

It is common for psychologists in specialized research and intervention areas (gossip, leadership, personality) to conclude that sapiens' EEA context best explains the patterns of their research, theory, and practice – as well as to extrapolate insights from near neighbors (other hominins, primates, elephants). For example, *Homo erectus* evolved to be very stable in its habitat ecosystem – sustaining for about 2.5 million years – partially because they achieved cooking food with fire and, it appears, pair bonding. Many complex, higher evolved mammals implicitly participate in group: information sharing, adaptation, learning, grooming, physical touch, conversing, enforcing moral codes, and supporting individual psychological recovery – such as mastering traumatic experiences and clearing them from the nervous system. Hominin bands, tribes, or clans in Pleistocene Africa were small (50-150 individuals), fundamentally egalitarian, and largely kin-based, with a semi-nomadic hunter-gatherer lifestyle. Before permanent settlements and agriculture, habitats were "well supplied with predators, poorly supplied with water, shelter, and food" (Vugt, Hogan, and Kaiser, 2008) – though humans still struggle with these hardships today. The sapiens' EEA includes all of these characteristics. Social organization qualities/characteristics in EEA groups are what diverse scientists increasingly realize our psychology and neurobiology still presume today.

Thus, many aspects of modern society are out-of-sync – i.e. not adequately harmonized – with our brain and psychology (our brain-mind-self).

Emerging evidence shows that dominance hierarchies, which are the norm in some primates, are not present in hominins and other primate species. On the one hand, with its evolutionarily near neighbor primates, sapiens share the push and pull of physical as well as nonverbal-vocal competitive testing, jousting, and ritual designed to be non-lethal (with specific exceptions) in inter-male competition.   On the other hand, alpha male domination with subjugation of the rest of the adults – as seen in some mammalian social species – is not human's natural psychology of leadership. Evidence is emerging that higher primates began to make democratic group decisions, pooling the experience of group members.   Macaques have recently been documented to exhibit democratic group behavior, following the majority to avoid splitting the group.

Patterns of dominance in the pre-'culture of modern humans' EEA are further socialized in sapiens into more egalitarian teamwork, give and take, sharing, verbal competition, and the corresponding moral codes and behaviors.   Any temporary hierarchy follows reverse dominance, as any temporary subordinates collaborate.   A democratic or consensual leadership style is the norm and natural way of thinking about, responding to, and shaping leaders.   "Fairness, integrity, competence, good judgment, generosity, humility, and concern for others" are valued. "Dominance and selfishness" is reacted to as the antithesis of leadership.   Leadership was distributed in situations and exercised "consensual, democratic, and transitory", based on areas of expertise and accumulated, proven trust.   (Van Vugt, Hogan, and Kaiser, 2008)

Psychology has documented that the ratio of positive to negative social interactions is healthy for individuals in the range from 2.9-to-one through 11-to-one (Fredrickson, B. L. and Losada, M., 2005).   This quality has likely been stable since humans evolved in the EEA and seems to be a universal psychological characteristic embedded in human nature.

In the EEA, one could also "do the math" and predict the differential participation of female and male sapiens in the more potentially violent activities.   Protecting, valuing, and nurturing women, not only enabled the group to have a new member every year or two, but assured the personal interactions with zero-to-three-year olds crucial for mature brain development (Schore, 1994).

Individuals biologically retain initiative and full individual organism species functioning, even as they grow sustained or embedded in attachments, family, tribe, and habitats and need social acceptance. All have survival value: "collective foraging and hunting, food sharing, division of labor, group defenses, and communal parenting provided a buffer against external threats"; "individual and group survival ... depended on cooperative effort and group cohesion." (Van Vugt, Hogan, and Kaiser, 2008)

Importantly, brain structures evolved to pattern non-social material physical pursuits and instinctual sexual/parenting behavior into more human objectives. Offspring and parents love each other. For adult sexual partners, sex turns into love and love desires touch, sex, and some amount of face-to-face companionship. For lovers, love comes to include touch, sex, and companionship.

Some earlier individual functions shifted into this particular type of social, such as:

a. Individual material success and acquisition became part of mutual and team success and acquisition; teaching cohorts and young included urges to personal best. Winning or shining in competition turned into non-lethal competition with relatively egalitarian distribution of the fruits of labor.
b. We and 'our relatively egalitarian team' needed material things and supplies – all need it and are on the same side; all got in and shared in the spoils.
c. Only-self-or-ego changed into self-in-attachment maturation to rapprochement, felt connection, and belonging bonds in tribes.
d. Instinctual mothering gave way to one-to-one attachments, family, and tribe child rearing and training.
e. Instinctual sex lead to a couple giving and receiving love, attachment, caring, companionship, and empathy as well as sex, which went along with shared childrearing and

family/group belonging.

With these crucial shifts, humans' strong and varied individual traits, talents, and capacities came to fruition. We could dedicate our skills to the success and resilience of the family/tribe in the face of challenges from various environments. For example, spending hours meticulously carving an arrowhead really only makes sense as a survival and reproductive strategy in a highly-cooperative and mutually beneficial context.

In this group context, individual contributions to defending, caring for, and/or enabling accomplishments of the tribe were routinely and practically closely interdependent with the same for self and family. Conversely, the group and family experienced survival and reproduction pressure to need and value that all or almost all individual persons can function at a high level, even as people competed and used their differences. Both too much and too little selfishness could be expected to be good enough integrated, managed, contained by external social processes and systems, with some assistance from natural world consequences. And all of significance were internalized or internally represented.

It is the coordinated evolution of upper inside and social outside that enables sapiens' cognitive and other higher cortical capacities and the emergence of persons who create culture. The pressures selecting for more and more varied upper locations of mirror neurons fit this frame. The 'deception capacities of mammalian species' payoff, overall, even less in uniquely human attachments, families, and tribes as compared to previous species. Face-to-face relating, the ability to name names and verbally explain details, and the egalitarian economic interdependence are powerful counterbalancing forces. This context further provides the prolonged childhood relating, nurturing, protecting, and training necessary for full maturation.

With more highly evolved young mammals having longer and more helpless stages of growth, solitary species (with instinctual/hormonal maternal care systems) transform to species, which include attachment and group surviving and reproducing relationships.

Walking upright required more helpless infants and longer childhoods. Babies needed to be born sooner in gestation so their heads would not be larger than the birth canal of their mom walking upright. As we have seen, the emerging self and cortex powers required not only even longer childhoods, but also highly individualized, secure personal attachments with their adults. Matured to an adult brain, the lower-upper lives in attachment to its upper self and to its attachment others (organized as a family) now including the adult's offspring.

Thinking of modern societies, cultures and lifestyles, it is hard to believe that such high-maintenance features in offspring could actually have been part of a winning formula that not only occurred, but was sustained over tens of thousands of years, enabling humans to dominate the natural world.

Human young represent their attachments, families, close relationships, and tribes, coded as people on their side. From the evolutionary perspective, selection pressures favored young who not only initially need, but implicitly expect that various parents and tribal elders, leaders, or authorities are doing things "for their good" and have their and the group's best interest at heart, enabling them to survive and thrive in their EEA as they accomplish the capacities and competencies and upper encode self, other, people, and the world "worldviews" (S. Epstein), which enable a healthy and satisfying adulthood-or not.

Individual parents and offspring comparably evolved to need, expect, and get involved so they, their attachments, family, and tribe (team) successfully, sustainably, and resiliently occupy and work habitat home and marketplace/economic niches which provide the base – the family home and group/economic community – necessary for the long process of raising, training, and educating children (learning from adult models, information, ideas, understandings, values), who themselves then have viable opportunities to accomplish and sustain the same. Individuals, couples, and multi-generational families further depend on each other for cooperation and live by norms of reciprocity, fairness, and returning favors in kind.

Further, just as mammals code a template of the body matched and corresponding to its involvement with the physical habitat, so do they code a conclusion or implicit identity template of who one is vis-à-vis what one experiences in relationships and interactions with others. Mixed-in with your thinking, your self contains copies of other people's past responses and messages and templates of yourself in relation to them. Your self has already coded the automatic body, emotion, and belief reactions you made during these encounters, as well as your subsequent automatic coping strategies and conclusions.

Nine

Experiential and Rational/Analytical Systems of Processing Information

As S. Epstein reports, in the televised program, a psychologist – without success – was aiming to keep the interchange between two rapists and a group of women "reasonable and productive". Many in the group kept focusing on the horrible crimes of the rapists and insisting on punishing them with the death penalty. Attempts by others to focus on what enables rapists to stop raping did not connect at all with the first set.

During our lives, humans use different systems for processing both internal and external information. We have one pattern of 'thinking' with a system of unconscious associating and processing that evolved up the phylogenetic scale – the experiential, associational system of processing information. That system's primary function is to learn from/to past, present, and anticipated future experience. The other is a rational/analytical system based in verbal reasoning. The two operate in parallel and are interactive.

Fifty plus years of research by psychologist S. Epstein and colleagues resulted in the synthesis of his Cognitive Experiential Self Theory (CEST). I consider this a huge psychological discovery. We now know the throughway for how humans think experientially, including when participating in the lower-upper construction method. Such an experiential-associative mode interacts with our rational/analytical-thinking mode. The experiential-associative-intuitive system of information processing integrates the cognitive and the psychodynamic unconscious and fully accounts for intuition. CEST, a dual process theory of personality, differentiates and formulates the rules, attributes, and interactions of two systems by which people process information.

CEST's formulations are brain-mind-self understandings, in my view. The following table shows a comparison of the two systems' qualities.

Table 8:

COMPARISON OF THE OPERATING PRINCIPLES AND ATTRIBUTES OF THE EXPERIENTIAL/INTUITIVE AND RATIONAL/ANALYTIC SYSTEMS

EXPERIENTIAL/INTUITIVE SYSTEM	RATIONAL/ANALYTIC SYSTEM
1. OPERATES BY AUTOMATICALLY LEARNING FROM EXPERIENCE	1. OPERATES BY CONSCIOUS REASONING
2. EMOTIONAL	2. AFFECT-FREE
3. MOTIVATED BY HEDONIC PRINCIPLE TO MAXIMIZE PLEASURE & MINIMIZE PAIN	3. MOTIVATED BY REALITY PRINCIPLE TO CONSTRUCT A REALISTIC, COHERENT MODEL OF THE WORLD
4. ASSOCIATIVE CONNECTIONS BETWEEN STIMULI, RESPONSES, & OUTOMES	4. CAUSE-&-EFFECT RELATIONS BETWEEN STIMULI, RESPONSES, & OUTCOMES

EXPERIENTIAL/INTUITIVE SYSTEM	RATIONAL/ANALYTIC SYSTEM
5. BEHAVIOR MEDIATED BY AUTOMATIC APPRAISAL OF EVENTS & "VIBES" FROM PAST RELEVANT EXPERIENCE	5. BEHAVIOR MEDIATED BY CONSCIOUS APPRAISAL OF EVENTS & OF POTENTIAL RESPONSES
6. NONVERBAL: ENCODES INFORMATION IN IMAGES, METAPHORS, SCENAROS, & NARRATIVES	6. VERBAL: ENCODES INFORMATION IN ABSTRACT SYMBOLS, WORDS, & NUMBERS
7. HOLISTIC	7. ANALYTIC
8. EFFORTLESS & MINIMALLY DEMANDING OF COGNITIVE RESOURCES	8. RELATIVELY EFFORTFUL & DEMANDING OF COGNITIVE RESOURCES
9. MORE RAPID PROCESSING: ORIENTED TOWARD IMMEDIATE ACTION	9. SLOWER PROCESSING: ORIENTED ALSO TOWARD DELAYED ACTION
10. RESISTANT TO CHANGE: CHANGES WITH REPETITIVE OR INTENSE EXPERIENCE	10. CHANGES MORE READILY: CHANGES WITH SPEED OF THOUGHT
11. MORE CRUDELY DIFFERENTIATED: BROAD GENERALIZATION GRADIENT; CATEGORICAL THINKING	11. MORE HIGHLY DIFFERENTIATED; DIMENSIONAL & NUANCED
12. MORE CRUDELY INTEGRATED: CONTEXT SPECIFIC; ORGANIZED BY COGNITIVE-AFFECTIVE NETWORKS	12. MORE HIGHLY INTEGRATED; ORGANIZED BY CONTEXT-GENERAL PRINCIPLES

EXPERIENTIAL/INTUITIVE SYSTEM	RATIONAL/ANALYTIC SYSTEM
13. EXPERIENCED PASSIVELY AND PRECONSCIOUSLY: WE ARE SEIZED BY OUR EMOTIONS & HAVE UNCONTROLLED SPONTANEOUS THOUGHTS	**13. EXPERIENCED ACTIVELY AND CONSCIOUSLY: WE BELIEVE WE ARE IN CONTROL OF OUR REASONING**
14. SELF-EVIDENTLY VALID: EXPERIENCING IS BELIEVING	**14. REQUIRES JUSTIFICATION VIA LOGIC & EVIDENCE**

Source: Epstein, S. (in press, 2010). Demystifying Intuition: What it is, What it Does, And How it Does it. University of Massachusetts at Amherst. Reprinted by permission.

The experiential is an adaptive system that starts lower, automatically associates from experience, is minimally demanding of cognitive resources, and evolved up the phylogenetic scale. The rational/analytical information processing system first appeared in humans.   Noteworthy, with its understanding the attributes and interaction of our two systems of information processing, CEST – an experimental- and empirically-derived, psychodynamic theory – integrated early-evolution 'thinking' with 'latest-in-evolution' capacities.   The experiential-associative-intuitive system of information processing also integrates the cognitive and the psychodynamic unconscious and fully accounts for intuition. Such feats are a testament to the latest-in-evolution intellectual powers of us humans as well as our shared cultural heritage and not-face-to-face teamwork.

CEST and Brain-Mind-Self

In my brain-mind-self framework, I differentially think of the experiential and the rational/analytical systems when keeping track of upper/lower and left/right cortical functions and capacities. While all of our brain uses the experiential system, only upper operates the rational/analytical system and the highest, most intelligent versions of the experiential processing. Lower responses are based purely on experiential associations.

Automatically and associatively learning from motivations, emotions, and body experiences employs, in humans, our experiential system. States like fear, hunger, or unmet basic social needs involve lower activations and responses.

With my brain-mind-self framework, any given person has the option to understand lower aspects of self-functioning as parts of self – and the further option to make sense of them as more childlike or more adult appetites, physical needs, and emotions. Understanding our two systems of processing information gives adults better conceptual frameworks, with which to formulate responses to the needs of self parts. In that regard, both frames implicitly postulate a universal best adult self-potential or capacity to utilize the combination of the two systems that enables what is healthy and adaptive for the circumstances and goals.

Remember the example of the white rats in psychology experiments whose payoffs were arranged by the scientists? Well, this lower and earlier-in-evolution responsiveness to the contingencies of payoff (i.e., conditioning) operates with the associative, experiential processing of CEST, as do we, when our cheese is moved.

These understandings fit with brain-mind-self's postulation of a lower-upper construction method. Scientists now understand such conditioning to be more than "a mechanical connection of stimuli with each other and with responses, but rather involves the construction of a model of the world that allows animals (including humans) to negotiate their environments adaptively (e.g., Hollis, 1997; Rescorla, 1988)" (Epstein, 2010). That 'model of the world' is the upper constructed by lower, which now guides lower functions.

CEST proposes that learning from examples, models, or observing can be grouped with conditioning. They have in common that they are automatic learning. By informing us that all are not just 'learning from experience', but are learning 'with our experiential system' gives us useful information about how our self is working with the experience and, in turn, what we may have to internally address, should we wish to shift our self thinking/feeling.

Brain-mind-self framework notes that the (mostly) upper located 'mirror' neurons can add more upper components to this observational learning. Our brains are filled with what scientists call 'mirror' neurons, which activate in us the same brain and body experience happening in someone whom we are observing. We can learn from our internal encounter with another's experiences, not just our own. This gives us more upper contributions to what lower learns.

Epstein and colleagues research finds that adults "respond to most situations in everyday life, particularly those that involve interpersonal relationships, primarily according to what they have automatically learned from past experience, i.e., by their experiential processing." This understanding enables persons to factor in the rules and attributes of the experiential system, when understanding the reactions to people and social situations.

Whether or not our rational/analytical system of information processing will be determined to be anatomically based more with left cortical powers or best considered the integrated processing of our two hemispheres, I find it very useful to realize that many intellectual products (reading the results of investigative reports, systematic conscious observing, reporting, science, journalism, law, etc.) involve both left hemisphere intellectual abilities and rational/analytic thinking.

Brain-Mind-Self Adaptability and the Experiential System of Information Processing

How normal and natural it is for each of us to have experiential learning start to take hold when we are experiencing something of self significance (e.g. involving livelihood, relationships, sense of self worth, acceptance/rejection, potentially threatening circumstances) is not extensively realized. When two things are associated together or a chain of events occurs, our experiential brain starts to link them up. "Neurons that fire together, wire together" (Hebb, 1949).

This is part of life. Individuals would benefits from factoring in this normal response mode. Having a personal time and place to explore an experience gives your brain a forum for it to 'have a dialogue' with itself, involving both information processing systems attending to the same event. Many times such self-conversations result in individual's correcting their own self experiential learning.

Another combination of brain-mind-self framework and CEST concerns that individuals consider that their experiential learnings and self conclusions many times have a lived internal self presence. They can be understood and given readapting or reprocessing opportunities.

The human brain, with its realistic sense that it is safe, secure, and in a different secure set of circumstances now and with its logical/analytical system of information processing, can reprocess experiential learning in certain psychotherapeutic contexts. Selecting specifics of, and combining, such components as

* Upper contributions
* Outer experiences (social, sensory, options for safety, choice, and appropriate self responsibility)
* Rational/analytical
* Highest right side guided adaptability/neuroplasticity
* Alternating bilateral sensory stimulation

allows us to readapt experiential learning much quicker and more efficiently.

There are exciting possibilities for applications to emotional education in this realm.

Ten

Self Adaptations, Living Memories

The first generation of neuroengineering computer models of the brain (artificial intelligence or AI) was a failure. The results suggested that it was impossible for the human brain to be performing all the functions it was, with its mere 100 million neurons. Said Kwabena Boahen of the University of Pennsylvania Neuroengineering Research Laboratory, back in the 1990s: "We now have microscopes that can see individual connections between brain neurons. They show that the brain can retract connections and make new ones in minutes. The brain deals with complexity by wiring itself up on the fly, based on the activity going on around it."

The brain being able to "wire itself up on the fly" (i.e. instantaneously reorganize its neural firings) creates ever-new iterations (instances) of the same brain. Each iteration is a copy of itself in terms of shape and basic structure, but the actual patterns of neural activity and component functioning (such as information processing and lower-upper construction) change every time.

Thus, neurons are neither changing randomly nor mechanically. Rather their exquisite choreography is based on your internal rhythms, growth, adaptations, and goals. Everything already in your brain – all your existing levels, experiences, and conclusions – will determine how your brain activates when something in outer triggers a new iteration. At the micro-scale, brain scans reveal neurons normally change their connections and firing patterns based on the situation and context to revamp what information they process and how.

To illustrate a circumstance where your brain instantly reorganizes: Have you ever walked into another room to get something, but once you got there, forgot what it was you wanted? So you go back to the previous room to remember? That technique works because your brain can switch iterations so quickly.

There are many ways to understand and scientifically study how our brain reorganizes and changes itself. Here we attend to brain-mind-self meanings. Activation of and communication among any of the universal functions and capacities thus far described in this book is occurring. Our experiential learnings and living memories (with the experiential-associative system of information processing described in the previous chapter) routinely activate. As presented previously, psychological science has documented that we come to have built-in self-wiring, which includes many different automatics of being (which include parts of self and self adaptations), even as we are designed to function as one. We live and activate these parts dynamically - meaning various parts of the brain communicate, coordinate, and integrate with each other in the process of orchestrating the changing activation of a healthy and successful life.

More About Self Adaptation To Significant Encounters

In Chapter Three, we learned: "Whenever our self assesses that we had a significant experience, we not only remember it, we neurologically make an internal copy of the whole thing. Whatever of your capacities and functions were operational, activated, and participating at the time (such as your fear circuit) and whatever was done to you (let's say you were attacked by a family member) become a part of your self that reenacts the experience every time it activates. The part of you whose job it was to make the copy gets neurologically 'topped' by its creation, thus becoming a functionally lower neural network."

In regards to self adaptations, our conscious self (see next chapter) may or may not:

- ♦ Agree with their messages
- ♦ Like them
- ♦ Identify with them
- ♦ Be loyal/attached to them
- ♦ Feel they are you
- ♦ Be aware of them

At any given point, lower and whole you has many existing lower-upper arrangements. Some of them can be considered a part of self and all have the job of representing some significant experience of yours – complete with the persons, places, and things involved, to a lesser or greater complexity. Thus, you have quite the menagerie as you go about living and pursuing work, love, happiness, and success. We have chains of activations – orchestrated dramas of parts of self communicating, needing, helping, protecting, coping, etc., that are as complex as anything humans are able to write and dramatize.

In addition to the presence of self adaptations themselves, they generate internal emotional and body reactions just as if we were now in the original encounter wherein we constructed them. We also then have coping reactions and defenses as we manage our internal self adaptations. All of these internal states can contribute to self strength, to increasing adaptability and resilience, or to problems.

Previous chapters explained that not just any social and physical environment – 'persons, places, and things' – would do. In child development, the self has to grow a right-sided self structure capable of both solo- (independent) and interdependent-self regulation and functioning. Social organization and interpersonal processes have to interact well-enough with the child as well as provide physical safety/security sufficient for self structure to develop.

When an individual experiences such constructions as supportive of his/her development and accomplishment, they integrate and harmonize within, increasing the individual's repertoire of competencies, experience, and so forth. When self adaptations involve family and friends 'good-enough' on our side, they are experienced as just us. All such adaptations just healthily and seamlessly become part of our self-functioning. Our particular wiring contains neural network configurations that enable us to 'be all we can be'.

These self adaptations – and our reactions to them – contribute to our uniqueness, personalities, competencies, capabilities. We have more flexibility, options, practical/adaptive knowledge, ability to handle more situations, more skills, more competent and effective member of our group, contributor; make decisions, take care of things, resolve things, mediate between people. Ideally we have a full community of personal internal protectors, nurturers, consultants, coaches, and managers – effectively working for us. They are accessible and support and enable us to shift and transform toward our desired future.

Lacking these leaves us vulnerable to health symptoms and problems and to a harder time t optimally achieving our goals. If others/events are treating us badly, either before we are developed enough to keep a boundary to the bad experience or when we are powerless to control the experience, we grow internal copies of them treating us badly. Unhealed traumas may become part of a self adaptation, which has the potential to be reactivated when current encounters are too similar, in their self meaning, to the past trauma.

Such self adaptations do not fit well-enough with our self development nature/needs. Hence, we have to work out new adaptations with them.

For some, this unhealed trauma may include events when their self was still developing in attachment. Developing self-structure involves numerous and complex configurations of self adaptations. As neurological architecture matures, the brain was working on integrating its varied elements and differentiating a separate psychological self from that of the parents. Self mal-adaptations and childhood trauma may interfere with gaining self-regulatory abilities and being able to alternate between healthily functioning solo or as part of a relationship/team, based on the circumstances.

Being a robust system, our brain-as-self monitors our environment and body and makes adjustments along the way to finding healthy, happy, and successful adult functioning. Our brain continuously and automatically works/learns to distinguish the lousy, unsafe, or not effective from the healthy, good, safe, or effective. This endeavor is not 'work' in the difficult sense, for your brain enjoys what it is doing.

Modern science and psychotherapy is piecing together the self and social processes that harness our natural powers to strengthen our self, meet developmental needs now that were not met when mal-adaptive self adaptations formed. We can experientially and adaptively process the traumas, losses, hurts, and other self blows to truly have them put away in the left hemisphere as just fact memories.

Eleven

Left Hemisphere Self,
Latest-in-Evolution Objective Interest/Capacities/Goals

Jarrod and Jessica are two year old twins, who are in the process of figuring out that mother does not know where they are if they do not tell or show her. To test what to them is still an unproven hypothesis, they hide in the bedroom closet. Mom comes to check on them and they are nowhere to be found. She calls and screams for them. Panicking, she runs outside looking for them. She starts to wonder if they have been kidnapped. All of a sudden the closet door bursts open and the twins fly triumphantly out, so proud of the success of their experiment.

Jarrod and Jessica can now internally observe and distinguish their internal representation of mom from mom herself. Is this latest-in-evolution self- differentiation (sense of selfhood) capacity of right hemisphere made easier by a differentiating left hemisphere? Probably.

A. Differentiation Of Our Left And Right Hemispheres

In the fierce competition for space in the cortex, some scientists noticed clues that the evolutionary advantage went to individual brains, which began to specialize existing necessary functions on one side, to free up the other side for additional capacities. One line of scientific reasoning notices that the 'heavy lifting' by the right-side – doing self, regulatory, and personal attachment/ relationship functioning integrated with the lower/whole – frees the left cortex to embody more of the latest in evolution capacities sapiens have.

Crows provide evidence (as well as non- *Homo sapiens* hominins) suggesting that the highly lateralized brain in vertebrates is what supports skill and alacrity in tool making (wherein we are most advanced). Humans are further interested in and motivated to make exact copies of a tool, to take the time to perfect the objects, to teach and verbally explain the craft and ideas to peers and young, with the goal of changing the habitat to control as well as benefit themselves.

Homo sapiens evolved at a time of dramatic climate change, when adding being able to communicate complex intentions, plans and strategies among your survival team proved to be a survival and rearing advantage. With one's bonded group and attachments already securely embedded in the individual heart, mind, and brain, more success went to those whose communication did not have to be in plain sight of each other. Our capacity for complex grammatical language is left. Corresponding with the development of an intricately wired speech region, adults with linguistic structure capabilities can give each other detailed information about what is happening in the habitat at places another is not or has not been.

B. Left Cognitive Capacities

Some of our more easily recognized latest-in-evolution qualities are on the left. Recall the left list from Chapter Two of our more cognitive left capacities. Notice many are the more intellectual, fact-focused, and schooled ones as well as about knowledge 'facts' of culture and history.

Remember the left cortex powers listed in **Table 3** of Chapter Two? Well, here they are again:

Table 9: Common Upper/Cortex Left-Sided
Psychological Processes

<u>Left Cortex</u>

• **Linear Reasoning, Inductive
Logic, Detail Focus**

• **Verbal Language, Naming,
Labeling**

• **Alphabet, Reading Language,
Writing**

• **Math, Counting, Arithmetic**

• Thinking/Reasoning about Ideas

• **Middle Range of Emotions
Processing** (not extremely
positive or negative)

• Sense of Will or Agency,
Choice, Pushing-Oneself

• **Social Self-Presentation**

• Interpreter

• **Conscious Awareness
Experienced as Separate from
Body and Emotions**

• Coherent Narrative Discourse

• **Word and Spoken Memories**

• **Fact-like Memories**

C. Left Hemisphere Capacities Overlap with Rational/Analytic
Information Processing

Our rational/anlaytic system is a language based reasoning
system. Gazzaniga's split brain research reveals such findings as
- ♦   the left hemisphere's need to form hpotheses, i.e., to
    figure out causality. research on the
- ♦   Left hemisphere does not miss the right hemisphere with
    its dominance for understadign the intentions of others.

I reproduced a portion of **Table 8** from Chapter Nine below.

Table 10:
**OPERATING PRINCIPLES AND ATTRIBUTES OF THE
RATIONAL/ANALYTIC SYSTEMS**

RATIONAL/ANALYTIC SYSTEM
1. OPERATES BY CONSCIOUS REASONING
2. AFFECT-FREE
3. MOTIVATED BY REALITY PRINCIPLE TO CONSTRUCT A REALISTIC, COHERENT MODEL OF THE WORLD
4. CAUSE-&-EFFECT RELATIONS BETWEEN STIMULI, RESPONSES, & OUTCOMES
5. BEHAVIOR MEDIATED BY CONSCIOUS APPRAISAL OF EVENTS & OF POTENTIAL RESPONSES

6. VERBAL: ENCODES INFORMATION IN ABSTRACT SYMBOLS, WORDS, & NUMBERS

7. ANALYTIC

8. RELATIVELY EFFORTFUL & DEMANDING OF COGNITIVE RESOURCES

9. SLOWER PROCESSING: ORIENTED ALSO TOWARD DELAYED ACTION

10. CHANGES MORE READILY: CHANGES WITH SPEED OF THOUGHT

11. MORE HIGHLY DIFFERENTIATED; DIMENSIONAL & NUANCED

12. MORE HIGHLY INTEGRATED; ORGANIZED BY CONTEXT-GENERAL PRINCIPLES

13. EXPERIENCED ACTIVELY AND CONSCIOUSLY: WE BELIEVE WE ARE IN CONTROL OF OUR REASONING

14. REQUIRES JUSTIFICATION VIA LOGIC & EVIDENCE

Source: Epstein, S. (in press, 2010). Demystifying Intuition: What it is, What it Does, And How it Does it. University of Massachusetts at Amherst. Reprinted by permission.

D. In being your self and dealing with life, consider your left powers your left Self.

Examples:

The left focuses on interpretation of events and works to assimilate perceived information into a comprehensive whole. The left has greatly increased capacity to at least temporally isolate to rational analytical thinking, planning, calculating, and focus on details. It is full of facts and information about topics the person is interested or schooled in.

The developed left seems to be able to direct the whole us in internal suppression and redirection of activation, but some researchers give evidence that this is due to the right hemisphere (through the pfc) enabling the left to do so.

The left needs accurate current and past information from the right and the outside or it will make up reasons (or be vulnerable to misinformation) explaining what it is noticing.

Memories deemed to be relevant to survival and thriving are kept stored 'activation ready' on the right side/subcortical, ready to be instantly and associatively activated with our experiential processing system. Living memories that the brain 'processes through' and concludes are over are put away as left-side fact memory.

Left-sided calculation includes the capacity to consider what is smart vs. stupid for the individual, regardless of the person's values, social loyalties, and short-term opportunities.

Split-brain research indicates our left side has a specialized Interpreter (as Gazzaniga, 2008, named it) of cause-and- effect function, one that could be out of internal touch with what the right side knows to be true or knew how to do. In other words, our left side could make it up.

My left will labor hard during the day to construct as precise a written narrative of what I know about the self and science topic I am addressing. I will then sleep on it. The next morning as I'm waking up, my mind is flowing with an elaborated rendition of what was so hard the night before.

My right has taken the bare bones and turned it into a flourishing cornucopia of big picture, interrelations among complex and varied parts, and so forth. Our right hemisphere wants to know what left wants it to work on. It will also observe what you are doing to figure out what to help with. That's what happened for Mario in the airplane.

Additionally we can explicitly ask our right hemisphere to help us out with something. Prayers and meditations, which connect to our deepest wisdom and guidance, are calling upon our right hemisphere's help.   Our right can supplement left with comprehensive understanding, intuition, imagination, and spatial skills.

E. Latest-in-Evolution Objective Interest/Capacities, Major Added Self Capacities

Latest-in-evolution functions and capacities, our highest intellect and sensibilities:

Whether left or right, we accomplished them related to the extreme differentiation of our hemispheres, and each presumes the two sides are working together.

The additional cortex space due to extreme lateralization facilitated an explosion of the ability to internally produce ideas including inventive ideas, then conscious ideas, and coherent personal and historical narratives, folk histories, and stories.

These latest-in-evolution capacities embody **our third type of responsiveness to the habitat**, to the thing/object world, including each other as objects of observation/experimentation. Individuals and their social group's greatly expanded motivation, interests, and abilities in deconstructing, observing, scientific or other experimenting ("let's see what works"), keeping databases, deciphering, and other changing of the physical habitat in the EEA to control and rearrange its conditions, functioning, and consequences to benefit themselves. We mastered the craft and technology skills of which many of us are familiar.

These unique interest/responsiveness aspects of the embodied human brain are observable in the preschooler. The left hemisphere is on board. The cortical maturation of the left and right hemispheres has each had a dominant initial-growth phase. Faster processing of the sapiens' cortex (compared to less evolved primates) comes online. A large pastime of preschool and school age sapiens children is observing and experimenting, as they figure out and master Piaget's cognitive stages of cause and effect.

Many a three year old has been the focus of their caregiver's ire for testing their newly realized ideas and abilities, e.g., destructively interacting with some home furnishing – such as cutting the bedspread with the scissors to check out its properties and powers.

Human structuring and planning of life projects, procedures, and processes, draws upon the integration of new capacities. Besides tool making, the human tribe uses its cause and effect thinking and rational/analytical thinking to plan and formulate the procedures, processes, and "rules of the game" for organizing social events and life staging: gather, hunt, store, cook food, retrieve water, bed down for the night, birth, and raise and train/educate the young, as is done in modern day societies, economies, and sports. All variety of individuals in the tribe are securely enough connected, shared with, valued, individuated, and humanized to organize events where some persons may be featured and fed without laboring, while others do the ready the food for consumption or provide the supplies of survival. Support and care giving are valued as important roles.

Humans further developed the capacity to step out of the drama of living in attachments, groups and habitats and not only observe, but judge and evaluate, like or not like it, try to figure it out (why, cause and effect), talk about it, receive observations from others about it, and so forth. We can even 'hang up on it' and deny something is even happening. On the other hand, we can also consciously contact, feel and talk to our very self-adaptations, getting to know points of view. We can observe or experience. You, as a living self, have capacities to:

* Stay separate as you talk about and observe yourself (observer);
* Connect with who/what you are and any felt experiences, capacities, or memories your body knows are relevant to you and your desired future; or
* Some combined and orchestrated dance, encounter, or alternating among the two.

With our sense of selfhood on the right and our social presentation skills on the left, we have choices regarding disclosure.

F. Three Options: Self Conscious Presence and Upper Self Agency or Intentionality Can be Left, Right, or Both

Science proposes that consciousness, our experience of waking awareness (which necessitates a cortex), can be right, left, or both (D. Seigel, 2001). Several lines of research document that 'hanging out' consciously on the left side is more associated with such states as the positive and calm in meditation and not being aware of negative emotions and appetites (even as the brain scan reveals such lower activation). Buddhist monks show activation more on the left side of the cortex (the pfc) as they report they are just observing what is occurring internally, staying out of emotions and thoughts.

Siegel proposes and classic split-brain research (Gazzaniga) has documented that we live self presence – and intelligence – in left and right. Many of these powers - left, right, or both – conscious attention, awareness, and control – are the powers of us being our pfc.

With our complexity, the next chapter makes the brief, but important point that we are designed to function as one and have our prevailing self configuration built with our specifics.

Twelve

Designed To Function As One:
Which Self Configurations Prevail

We, the whole, are designed to dynamically function as one. We have prevailing self configurations – the states, capacities, behavioral, procedural responses, self constructs of self/world/other that are dominant or take charge of whole now functioning. What prevails we subjectively experience as our mind, spirit, and behavior.

Which Self Configurations Prevail

Example 1: A Crash

Does right, lower, or left prevail when experiencing impact in an airplane crash?

A trauma specialist once told me that a person is statistically more likely to survive an airline crash if they do not know it is going to happen in advance. It seems that the brain is less likely to go into lower responses prevailing, such as fear and freeze responses taking over.

Apparently, when you don't consciously know a crash is imminent, your right hemisphere can optimally calculate the best position for you to be in as the plane crashes. Its calculations include how to best move each individual body limb based on how objects are moving in the vicinity and other people's reactions.

Knowing about the imminent crash makes it more likely that the person's lower will take over and switch into any of the 'unthinking' emotional reactions.   Scientists call this response a 'low road reaction', as opposed to a 'high road reaction' involving the cortices.

Not everyone who does not know about the coming crash will switch into the right brain taking charge. For some people, as soon as any part of their brain realizes a crash is about to happen, their lower immediately activates emotions and/or fight, flight, freeze, or 'help me' reactions (the vagal nerve; see Chapter Four).

Example 2: Five People Afraid of Heights

Each one is offered this chance: If they climb to the top of a 50-foot ladder, they will receive $70,000.00.

Who will do it?   What are the aftereffects?   These give us information about what prevails, in the short, immediate, and long term. Will s/he do it? How will each fare afterwards?

Scene: Because they are afraid of heights, we know lower fear circuits are activated. Because outer is offering a huge payoff, we know lower circuits seeking reward are also activated. Thus, the two lower activations are at cross-purposes.

How does upper get involved?

1. For each person who attempts the climb, we know the conscious upper/left has set a goal and the right is at least working to enable it.

2. For those who refuse, lower 'fear' has immediately trumped lower 'seeking reward'.

3. Watching one climber, you can almost see in his face his right hemisphere imagery visualizing the $70,000, spurring him on to overcome the fear and climb, despite sweating and panic.

4. As another climbs you can see a caving in happening as the person stops.

Afterwards, again there is variety in what prevails. Four out of five people do it.

1. Peter pushes down his fear temporarily and climbs up the ladder. While he is happy he won the money, afterwards he is badly shaken. It will take him a while to recover. He does, but the fear of heights remains. Doesn't make it better, but not worse. The money was worth it. Never again he says.

2. While Libby's heart was wildly pumping as she climbed, afterwards the fear seems gone for good. The exposure cleared the problem from her system. It's gone and stays gone. Not only was the left strong enough, but right participated to enable her to 'process' or 'work through' the fear.

3. Carla made it, but afterwards degenerates to increased phobia, panic attacks. She was re-traumatized. Her brain was unable to distinguish past bad experiences – living memories – from this one. Her right-sided self needs strengthening to turn things around.

4. Joe did not go on the ladder. A drunk close family member, whom he trusted, had left him too alone and unsupported too many times. Once, about a year old, he was left on the ledge of a counter. His upper map/conclusions of the world and people have generalized that most people and places are unsafe. He develops agoraphobia (he can not leave the house) and rarely leaves his parents basement. The fear has engulfed his whole. Those pre-prepared response circuits are chosen by his cerebellum.

5. Sharon is glad she did it. She decides to expose herself to further climbs to continue to adapt her lower to not be afraid. Her right and left uppers are working the lowers to stay calm when climbing, as the cerebellum chooses climbing as the option enacted.

Building internal self resources, drawing on core, strengthening right and left sides are all options available in psychotherapy and coaching that enable a person to readapt out of a phobia.

In any situation, something prevails overall. Your goal is to healthily and sustainably realize you. Before you learn of the brain basis of your repair and improvement powers, it is useful to know how some of human's basic themes are patterned in *Homo sapiens'* 'culture of modern humans'.

Thirteen

The Culture Crafted by 'Modern Humans'

With strong universal functions and capacities, humans engaged in tool making, craftsmanship, and trading with the goal of changing the habitat to control as well as benefit themselves. Growing our own food and domesticating animals, rather than hunting and gathering daily, freed some of us to not have to directly grow, hunt, or gather food. We both greatly increased our divisions of labor and enriched some individuals so they did not have to physically labor or labor at all. Some have become traders, scholars, artists, inventors, scientists, teachers, health providers, politicians, lawyers, stockbrokers, owners, stockholders, art and sport performers, writers, and more. We enjoy extensive material goods, services, and the fruits of trade and technology.

Much has been appreciated and written about the benefits and likewise for some of the broader costs such as mass violence, war, and sexual predation on our own as well as ecosystem destruction. Individual, family, and group histories tell the stories. The benefits have come with other unintended negative consequences and vulnerabilities, related to the brain-mind-self realm.

An unintended consequence regarding our interpersonal and social-organization processes occurred with our accomplishment of the 'culture of modern humans'. Our specific stable forms of interpersonal processes and organization, a key contributory cultural condition which enabled the very genetic evolution of *Homo sapiens,* is fundamentally disrupted now that we live adapted to farming and ranching/herding, sedentism, and large impersonal, non-face-to-face societies. One serious hypothesis to consider is that disrupted/ changed social conditions are significant contributors to high rates of mental problems and violence.

Now we have both one-on-one, small group, or micro level interactions, and societal, large group, or macro level interactions, whereas throughout our EEA before the 'culture of modern humans', the former small dyads, families, and group/tribe level was all that existed. This change in individuals' face-to-face relating is one difference between our EEA-formed expectations and our current reality.

Further, humans' brains expect modern-day organizational structures to involve the same group dynamics as the EEA. We evolved to presume that our group and our selves were on our side. How much we need each other and need grounding in our interpersonal relationships for health and well-being is known. It is vital to our mental health to be seen and valued as an individual. How much our biology presumes that our strong individual capacities are contained in our social bonds has been underestimated, even as we have the capacity for duplicity and insensitivity.

People can now function for each other as asocial-only habitat components, thing-like components, or objects – just a means to an end or a thing to be used. Before, such components became patterned in interactions between people inherently connected to others, between basically valued or bonded persons. This is the 'follow the money' adage. When individuals deal with an industry or business, they benefit from knowing the entity has its role of providing its service/product for the customer as well as its role to maximize money/payoff, the thing/object role.

Our higher asocial powers/capacities evolved to overall increase the survival/reproduction for all of our strong individuals (unlike wolves, for example, all our adults are biologically and socially developed to procreate). Now our powers can be used in ways that help, harm, or both – ourselves, others, and our groups. Lower asocial elements exist even in higher asocial pursuits.

For their own sanity and for greater success in their goals - ranging from successful business ventures, group and individual dynamic management, weight loss, creative and sports accomplishments, religious community, raising and educating children, to housing events – individuals have benefited from distinguishing whole person relating from the pursuit of only asocial motives in normal commerce.

This is in addition to knowing and improving your own brain-mind-self with our current, desired and expected future, given any needed repair regarding your history: your particular grown array of universal components and ways of activating are on board at a particular point in development.

Certain behaviors we see – selling bottled water at a disaster for an outrageous price; supervisors taking advantage of supervisees; patterns where rent increases precisely occur when there is an increase of foreclosed homes (both are supply and demand in disasters); lobbyists/ employees, employers using sex to get some advantage; community members of power/money getting out of the obligations/laws that apply to the rest of us (drunk drivers, not paying their bills); the health care industry executive/manager looking for ways to cut costs sees the fellow executive who has a chronically ill child as an irritant to be outmaneuvered – are examples of lower asocial/thing/object transacting. Like the Gorilla hunting parties, they are focused on the payoff. Other more human sensibilities are not attended to.

Psychologists have empirically documented how a confluence of external conditions, can trigger ordinary good people in civilized, democratic societies to abandon good moral judgment and resort to violence and oppression (Zimbardo, 2007 and Milgram, 1975). Vice versa, research results and applications articulating broad positive individual and social sensibilities – and their needed appropriately enabling conditions – (Fredrickson and Losada, 2005; Seligman, 1975; Gottman 1994) continue to be added in the 21st century.

Regarding our systematic social and economic organization, our brain-mind-self seems to not do that well with over-having, just as we don't do well with under-having. More shared and distributed having, work, and play in a social community is our species.

We are healthier and more resilient with short feedback loops, communication with, and providing of payoffs to actual people on the ground and who are doing the work. The extreme division and distribution of labor and payoffs invites mental problems in adults. When those adults are parents or authorities involved with children, the mental health of the children are then further damaged.

Thriving middle classes; codified rights, respect, wealth; and viable work/rest/play/family rhythms in society would make all of us happier. In common with the rat in the psychology laboratory, history is replete with examples of individuals and groups which have been captured or trapped in work roles, not valued or protected and vice versa the individuals, groups, and enterprises/industries, which captured them for purposes of gaining payoffs of various sorts, from precious metals to fertile soil or trade routes.

We depend on accurate and social/humanized information and alerting of perception (sense) & attention from relative equals concerning what we are doing. Example about the lower payoff motive: The founder of the Society for Primate Preservation was for years a hunter who participated in hunting parties sneaking up on gorilla tribes, thrilling in killing and stuffing payoffs. One day he looked up at the face of his stuffed ape in his home, saw the sentience and emotion, and realized they are like us, i.e., he empathized. His conscious attention, observation, or mindfulness changed internal focus to more humanized and social. He started videotaping the ape hunting parties and organized viewing parties for the hunters.   Their conscious attention now saw (sensed, perceived) their ambushes for the cruel slaughter they were. Almost all gave up such hunting and joined forces to protect and prevent such slaughter.

Another Unintended Consequence

Human's brains were selected to adapt to varying, even harsh, physical ecosystem habitats of wild nature. Our biology is not as useful in protecting us from threats caused collectively by ourselves.   We evolved to react to threats and loss more experienced as lower needs and responses. We are more likely to respond to threats that have characteristics to which our lower response repertoire reacts. If the threats are visible to our senses, immediate, with simple cause and effect (the tiger is going to gouge my neck and kill me) personally experienced by us or have historical/community precedent in our lifetime.   We biologically expect life-threatening social threats are caused by something or someone outside- not the family, an enemy, or another tribe. Finally, it is easier for us to assume that serious problems and difficulties are externally caused, not due to self-interfering internal components of individual or group.

The Human Journey

The story of brain-mind-self is part of a journey of humans, where history, science, and psychotherapy/health/performance practice are still exploring the terrain. Just as individual persons are still learning how to accomplish having security and a sense of individual autonomy, so is society still working on accomplishing that. The *Journal of Psychohistory* concluded that the history of childhood is a nightmare from which humanity is just beginning to recover (citation not yet located). Individual persons, families and groups are still learning how to accomplish key elements of resilient self and social health, how to best establish humanity and earth resilient & sustainable economic entities/businesses/corporations, resource stewardship, social interacting communities, raising and educating children, resilient individual health/well-being, and shared values.

Fourteen

Frontal Lobes, Pfc, and More Upper

Meet the sub-regions of our brain that play the unique, newest-in-evolution, most sophisticated, and largest controlling/directing upper roles. These latest-in-evolution aspects of certain brain areas allow us the vast range of our neuroplastic self improvement powers. While science will continue to advance and improve our understanding of just what part of the brain dose what, we can make use of the regions thus far proposed. Here I highlight major players:

1. The Front of the Cortex: Frontal Lobes
2. The prefrontal cortex (pfc)
3. The cerebellum
4. Newest parts of the brainstem
5. Possibly the thalamus (see 'More for Scientists' below)

In particular, the pfc, cerebellum, and the latest-in-evolution brainstem constitute a leadership/parenting 'team' in the dramatic production of our self. All such newest-in-evolution brainstem components (which includes the cerebellum, as it is actually an outgrowth of the brainstem, albeit scientifically studied as its own brain part) I consider to be part of upper – along with the pfc.

Our pfc takes most of the executive functions/roles: director, parent, responsible adult, and stage manager. The actors are our various pre-prepared response repertoires, including functions, capacities, and self adaptations (involving the scenes and participants extant at the time that you/self/brain made a particular adaptation).

The cerebellum makes the final 'behavioral production performance' decisions regarding staging, timing, choreography, costume, and so forth. Decisions about last minute emergencies, whether due to externally perceived threats or internally activated pre-prepared response repertoires, are the purview of the cerebellum.

The brainstem (due to its newest evolved attention roles and capabilities) has some veto power, including 'line item veto' power over where/when we pay attention – including going to sleep and waking up. It functions as the lighting crew and determines whether and where the light is on, shutting off and turning on specifics relating to alertness and consciousness. The brainstem teams up with its cerebellum for final curtain.

How is conscious you involved? Do you get to see who is there, where, doing what? Do you get to see the performance at all? Or only hear it? Is smell or peripheral vision turned off?

Example: A middle-aged man was watching a movie that included a story line and scenes too reminiscent of some childhood bad experiences. All of a sudden he went blind. His friends had to shepherd him out of the movie theatre. A couple of hours later it all cleared up. Extensive medical testing found no underlying medical problem. I suspect the brainstem – perhaps advised by the pfc to do so – decided to shield this man from being exposed to a scene too similar to his original trauma.

Frontal Lobes

Lumping left and right together, much of our upper capacities come from the front of the cortex, the frontal lobes, including:

♦   Internalized thought/imagery/language/models/beliefs;
♦   Social life and self/social development structure and models;

- Representing lower implicit biological rhythms and cycles (time, day, week, season, etc.) in concepts;
- Representing lower appetites in concepts;
- Capacity for flexibility and adaptation to changing conditions;
- Conscious power to direct inner and outer attention; and
- Working memory: internal control/change strategic powers, inner space construction.

This list shows some results of the lower-upper construction method as well as cortical contributions to the brain-mind-self's leadership/parenting team. The pfc is part of the cortex, with its specific roles the subject of the next section.

Prefrontal Cortex, Pfc

Much brain upper work is performed by an even smaller part of the front – the front of the frontal lobes – called the prefrontal cortex (pfc). Our left and right pfcs enable us to:

1. Shape bodily processes (oversee brainstem activity)
2. Pause before we act
3. Have insight
4. Empathize
5. Morally judge

Further, the left and right of the pfc do any upper integration work. Our pfcs help coordinate, balance, and integrate signals from all other areas/regions of the brain and body. Additionally, they link those signals to the sense signals we receive from and send to the external world – social and physical (including learned sensory and performance skills).

Our pfc constantly receives signals from, integrates, and manages our bodily and survival/reproduction motivational concerns, our emotional evaluations, and the pathways that connect us to the social world and minds of other brains.

Integration involves all the interested and relevant parties being present, accounted for, or part of the discussions.

In previous chapters, I primarily discuss our sided capacities, but so far this chapter I have not differentiated. To defend the importance of our right side, the functioning of our right cortex cannot be practically distinguished from pfc functioning. Our right-sided powers of self-in-attachment structure, sense of selfhood, moral judgment templates, and observing/mindsight/mindfulness all map and overlap with pfc functioning. Essentially, our right cortex enables the pfc to do what it does. In defense of the left, the right and pfc are at the service of left hemisphere us. They want to know what left hemisphere us wants done – for a particular task, day, or long-term goal.

Pfc Working Memory

Our left and right pfcs are the location of what scientists call working memory, an extremely versatile ability. In this section, I focus on working memory as an inner place/space where we work on whatever we want or whatever tasks we have that day.

A kind of internal control, management, and improvement room, your working memory is a place for the jobs of the day – a place that enables you to perform, learn, and pay attention – all brought to you by your pfc.

Other primate species may, in fact, have a larger working memory capacity than *Homo sapiens*, but our pfcs net us greater flexibility, adaptability, and change powers using our working memory. This means we can do any needed change or repair work on ourselves, as opposed to solely relying on others (as the reader saw in Gorilla example of Chapter Seven). We can guide and regulate our own self-functioning and its improvement toward goals we select in ways compatible with who and how we are.

On the left, we might think of an abstract idea, concept, judgment, or value; or we could decide on and plan for a conscious goal we'd like to achieve. On the right, we might imagine, fantasize, or elaborate a whole story concerning that same idea or goal. They come together and collaborate in our pfc's working memory. Further, our pfc enables us to bring such upper into lower – to feelings and body – should we choose to.

Humans can guide, manage, and really do the work of improving the rest of the brain from this teamwork and location. As much as the previous chapters discussed our development and adaptations from lower up, with our pfc team, we can reconstruct from outer and upper as well as lower. We engage functions, parts, and capacities from both sides in particular combinations of requested (conscious) and automatic (unconscious) – just one more example of the yin-yang whole mentioned in the Introduction.

Psychotherapy and coaching work involving working memory is amazing: The adult human brain – due to the addition of these capacities higher in the brain (the pfc), body, and evolution – can use new experiences to heal, rewire, and rebuild what is traumatized or inadequately constructed, as part of self-strength work. Many of the abilities we already use for interacting with other people and the world, you may also employ for interacting within yourself.

For instance, we pay attention. Guiding inner attention is popularly called by various names: mindfulness, mindful awareness, conscious awareness, observing presence, focused attention, and working memory. Attention powers can be employed internally to be self aware, conscious, and have mindsight. They can team up with other capacities to increase mental strength, heal, and repair.

Just as you can direct your attention to and interact with a person, place, or thing externally, you can direct your attention to and interact with a part of self, experience, or living memory internally.

Just as you can structure and organize your environment, such as by moving objects together or apart, you can structure and organize your mind by moving, connecting, or disconnecting parts of self, experiences, and living memories.

Just as you can parent, care for, like/dislike, guide, manage, and feel close to or distant from external others, you can parent, care for, like/dislike, and feel close to or distant from parts of self.

Just as you can govern, caretake, or provide services for groups of individuals (either working with the grassroots or acting very disconnected from them) externally, so too can you govern parts of self with this same variety internally.

Just as you can form images and animate image and creative processes concerning external things in your mind's eye, so can you likewise do with internal presentations of self components. The self can identify with and accept – or not – what is present.

We can put something in front of our mind; bring it forward – into our mind's eye. We can use and direct the attention of our conscious alertness.

To summarize, here in working memory resides our capacities to direct, produce, structure, place, connect, time, organize, negotiate, mediate, and engage – both between and among people and parts of self. To some extent, we have the ability to write fictional scripts/novels or perform made-up characters and assign staging in plays/movies because we already do those same roles for ourselves. We're transferring internal skills and processes into the world of acting, fiction, and make-believe.

The Cerebellum and the Newest-in-Evolution Brainstem
Modifications

Now as amazing as all this conscious attention and restructuring is, we have even more powers. The above is not the whole story of you directing and controlling your attention and behavior.

That big bulge on the back of the brainstem, the cerebellum, evolves and develops in tandem with the cortex, meaning evolutionarily as the cortex gets larger, so does the cerebellum. Torrents of information pour into the cerebellum for final processing and decision-making. Simultaneously, it sends out instructions to various other regions of the brain, telling them their job in the final output flow. Our cerebellum (with the received signals for which it is voting) can pull victory out of the jaws of defeat – or vice versa – in all varieties of social and physical performance. Optimal performance 'in the flow' can be orchestrated, as can accidents and missteps. Our highest right side intelligence 'figuring out' the successful way forward contributes to those last minute saves (like Mario in the airplane example of Chapter One). The cerebellum acts as the brain's clearinghouse and chooses which of right/pfc's last minute or emergency signals to let prevail and take control of your performance or behavioral enactment.

Moreover, we have parts of the brainstem that evolved to support interacting with the frontal lobes and pfc. Anesthesiologists turn our consciousness on and off to various degrees for a requested medical surgery. Sometimes our consciousness level can be brought to the equivalent of a hypnotic state, and it can even be turned off completely. Likewise, automatic you (including parts of your brainstem) controls an internal source of attention and sleepiness based on your goals and rhythms of alertness, sleep, and rest.

More Upper In Action

To give one of infinite real-world examples, the pfc, cerebellum, and brainstem work together to assess the best course of action in an emergency situation.

A man finishes cleaning his backyard swimming pool and walks around the edge to put away the skimmer. As he turns a corner of the pool, he missteps and suddenly feels his left foot go into the water, while his right stays on the cement. His first instinct is to fall into the water and avoid hitting the side of the pool (and in particular to protect his testicles from landing on the hard edge – an additional lower motivational message that made it into the cerebellum's final 'performance decision'), so he starts to push off with his right leg. However, gravity has other plans.

Moving his leg on the ground requires the neural signal to go from the cerebellum to the multilevel motor-action system (recall Chapter Five), then all the way down to his leg; there is simply not enough time – or potential energy. The man ends up crashing into the water facing, and more-or-less parallel with, the pool wall – passing his foot still on the cement on the way down. His right leg twisted like a towel and, broken in three places with mincemeat soft tissue, fell in with the rest of his body.

The man found himself soaking wet with an unusable leg, but clinging to the side of the pool (so only wet from the chest down). Apparently his cerebellum (and strong upper body muscles) was able to control his arm movements in time to catch the pool edge. In nanoseconds, his pfc assesses the situation and realizes he might drown if he blacked out from the pain with no one else around. So he uses his (brainstem) natural adrenalin rush plus heightened alertness to pull himself out of the pool and drag himself inside. He screams and his wife runs from the other end of the house to his aid.

As the leadership/parenting 'team' of our internal production, the pfc right and left sides, cerebellum, and latest-in-evolution brainstem together enable and conduct our flexibility, adaptability, and self neuroplasticity.   For instance, when a self adaptation activates (as discussed in Chapter Ten), the pfc can match firing patterns happening in the present to those that happened at the time when the triggered self adaptation first became a part of the person. The brain can then – in an instant – switch from present to the past inner movie, or some commingled version of the two, without the individual even being aware of past influences.

These brain centers give us the options of experiencing our selves as being the driver or the car (using the analogy first presented in the introduction). Humans have huge array of different self adaptations, models, constructs, and instructions guiding what we do. Our flexibility and adaptability can become too much, too many, too convoluted, and, in the extreme, we can be like a badly wired house in need of updating, streamlining, and pruning. It is not uncommon for the neural wiring within, between, and among our pre-prepared response repertoires – including functions, capacities, and self adaptations – to need a minor or major remodel. As we age, it is almost as if new owners periodically move in and add their own additional configurations, so that by adulthood, we may have grown quite a hodgepodge. Professional practitioners notice that self wiring structure can sometimes grow to be quite disorganized, dissociated, split, or disordered – affecting our very sense of selfhood or sanity.

Adult Self Neuroplasticity

With the pfc, cerebellum, and latest-in-evolution brainstem regions, we have the capacities necessary to participate as the co-mechanic facilitating our own self-wiring repair. Gazzaniga refers to this power as a kind of mind or downward causation.   These regions also enable us to adaptively take in bottom up contributions from the body, posture, breathing, and activity habits.

The end of Chapter Nine discussed some interesting applications regarding how individuals can normally communicate and reprocess everyday experiential learning.   Both current happenings and living memories could straightforwardly be reworked, if people understood the process and had some overview of human's normal construction in this manner. Such work uses our newest-in-evolution capacities.

Systematically using newer brain capacities/ parts in inner improvement and repair work enables you to healthily master what is interfering or troubling and gives you more productive options in how you get involved in the performances, challenges, and issues that matter to you.

Possible accomplishments:

* Calm, containment (closing up, putting away), grounding, centering;
* Accessing and strengthening positive top-down and bottom-up qualities and resources;
* Being present as your desired future 'soaks in' now;
* Knowledge, information, and perspective;
* Synergize capacities to increase mental/body strength;
* Internally process and remedy lousy experiences (neglect, unmet needs, injury, trauma, loss);
* Readapt maladaptive parts of self;
* Facilitate self-development and integration; and
* Strengthen inner regulatory capacities.

Your self system can transforms to:

* Experience inner peace, health, happiness, and calm as fits the situation,
* Integrate and manifest your real-self wisdom and feelings,
* Actualize your talents and competencies,

* Use creativity and solving challenges in your circumstances now, and
* Sustain satisfying relationships, while having a self.

More for Scientists

For those who are interested, I hold in mind the two very broad sensory information processing pathways running through the brain.

Information coming from the external senses comes in and up the brain's dorsal Thalamic-Neocortical Axis (the Somatic-Cognitive Nervous System of the brain). The **thalamus** accomplishes integration among parts of self and helps integrate the senses and contributes to linking and binding visual, auditory, and tactile sensory information, time, and cerebral cortex/cognitive elements. Thalamic functions contribute to the integration of information. They participate in consciousness, as they are the route from the brainstem sources of alertness and arousal for the stimulation of the cortex's perceptions, cognitions, evaluations, and beliefs.

Information coming from the internal (body) and chemical senses (except smell) come in and up the Hypothalamic-Limbic Axis, first processed in the brain's ventral Visceral- Emotive Nervous System. This is the subcortical location of the first brain information processing of physical needs, social needs, and needs to be, like, and grow your self. Included are our capacities for feelings, gut reactions, body presence, posture, and body states/physical functioning. Our pleasure/pain and natural feel-good neurochemical/hormone systems all root in this, our ventral Visceral-Emotional Nervous System.

On the way to generating behavioral output performed by the body, these two major sensory information processing systems converge on the sensory and motor control programs of the Basal Ganglia (a structure shared by all vertebrates from reptiles through human), called the Extrapyramidal Action Nervous System. This

Extrapyramidal Action nervous system is the primary brain seat of compulsive behavior problems.

The thalamus is one neural synapse away from the other axis just described. That visceral-emotional axis includes our emotional reaction brain organ, the amygdala, where such reactions as fear are triggered. This communication can contribute to emotional interference with consciousness and memory as well as the dispersal of feel-good hormones through out the brain. Finally it does contribute to neuroplasticity. Future research may very well propose that the thalamus is one of the leadership/parenting team for health and improvement. (J. Panksepp, 2008, U. Bergmann, 2010)

Fifteen

Normal and Necessary

Most people have little idea how normal and necessary it is to learn about and care for brain-mind-self as well as to participate in processing, adapting, and repairing self-experiences. Few of us know how to dance between equals, and that caring and trusted others are part of whom our brain presumes is on our side.

Yet all that is what enables our brain to successfully do its roles in maintaining and recovering physical and mental health and enables us to realize and perform our best.

We use it, lose it, and recover all the time. It is when we do not work with our brain to 'work it through', that we diminish our health. Like brushing our teeth, we save a lot of social embarrassment, physical pain, and money, if we learn what to build into a preventive routine.

Don't sweat it. You come equipped with amazing capacities for relating, lower-upper self processing, re-adaptation, recovery, and repair. All you have to do is learn how we humans are put together (the brain-mind-self framework can help), your particular inner and outer version (with your strengths, resources, and problems/blocks), and what you'd like to change, while being accepting of – and willing to work with – what you find. People can be caring and trusted co-guides with each other. That is all. (Smile.) Okay, sweat it. But also talk. There is a way to 'you realized'.

Image 5

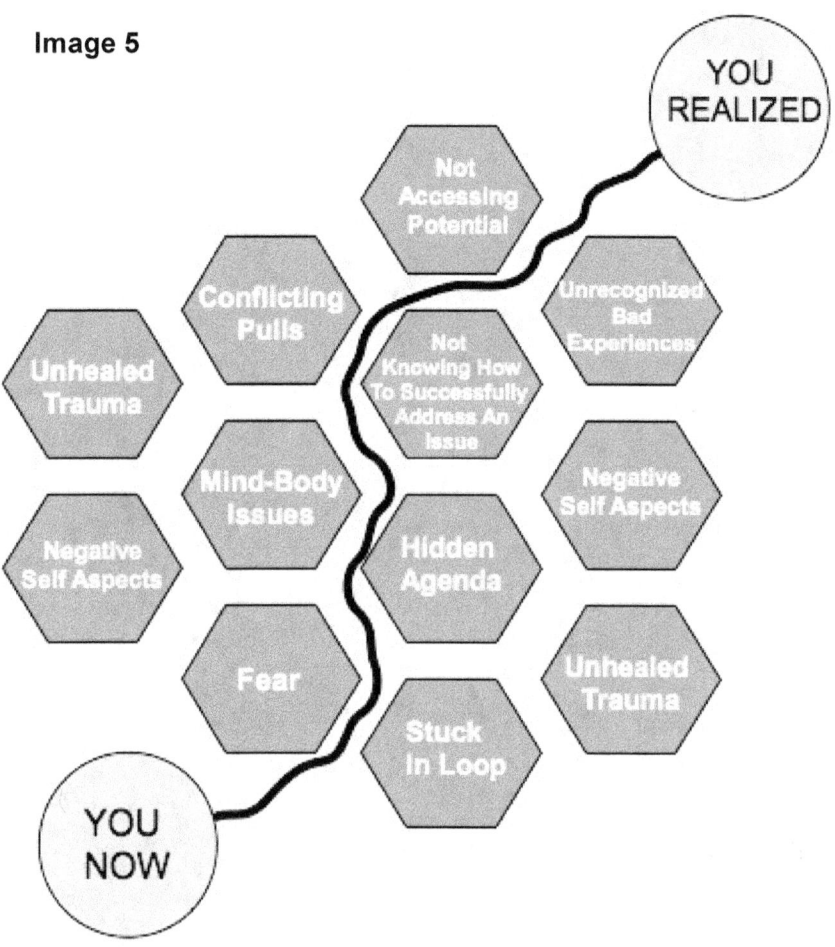

References

Ainsworth, M. D. S., Belhar, M. C., Waters, E., & Wallace, S. (1978). *Patterns of attachment: A psychological study of the strange situation.* Hillsdale, NJ: Lawrence Erlbaum.

Amini, F., Lannon, R., & Lewis, T. (2000). *A ganeral theroy of love.* Random House.

Andreasen, N. C. (2001). *Brave new brain: Conquering mental illness in the era of the genome.* New York: Oxford University Press.

Arden, JB., & Linford, L. (2009). *Brain-based therapy with adults: Evidence-based treatment for everyday practice.* Hoboken, NJ: Wiley & Sons.

Augustine, G. J., Chikaraishi, D. M., Ehlers, M. D., Einstein, G., Fitzpatrick, D., Hall, W. C., et al. (2004). *Neuroscience: Third Edition* (Third Edition). Sunderland, Massachusetts: Sinauer Assiciates,Inc.

Bainbridge, D. (2008). *Beyond the zonules of zinn: A fantastic journey through your brain.* Cambridge, MA: Harvard University Press.

Bargh, J. A., & Chartrand, T. L. (1999). The unbearable automaticity of being. *American Psychologist, 54,* 462-479.

Bergmann, U. (2010) EMDR's Neurobiological Mechanisms of Action: A Survey of 20 Years of Searching. *Journal of EMDR Practice and Research.* 4(1) 2010, 22-42.

Blumberg, M.S. (2009). *Freaks of nature: What anomalies tell us about development.* New York: Oxford University Press.

Bowelby, J. (1969). *Attachment and loss, Vol. 1: Attachment.* New York: Basic Books.

Bowelby, J. (1973). *Separation.* New York: Basic Books.

Bowelby, J. (1988). *A secure base. Clinical applications of attachment theory.* London: Routledge.

Breckler, S.J. (2006). The newest age of reductionism. *Monitor on Psychology,* 37(NO.8), 23.

Busch, R. M., McBride, A., Booth, J. E., & Duchnick, J. J. (2005). Role of executive functioning in verbal and visual memory. *Neuropsychology, Vol.19* (No.2), 171-180.

Casadaban, A.B. (2006). *A model of the mind of the brain to envelop psychotherapy integration.* [Los Angeles, CA, USA]. Paper presented at the meeting of Society for the Exploration of Psychotherapy Integration (SEPI).

Cozolino, L. (2006). *The neuroscience of human relationships: attachment and the developing social brain.* New York: W.W. Norton & Co.

Cozolino,L. (2002). *The neuroscience of psychotherapy: Building and rebuilding of the human brain.* New York: Norton.

Damasio, A. R. (1999). *The feeling of what happens: Body and emotion in the making of consciousness.* New York: Harcourt Brace.

Davis, M. (2006). Neural systems involved in fear and anxiety measured with fear-potentiated startle. *American Psychologist, Vol 61*(No.8), 738-741.

Diamond, J. (1999). *Guns. germs, and steel: The fates of human societies.* New York: W W Norton and Co.

Diamond, J. (2005) *Collapse: How societies choose to fail or succeed.* England: Viking.

Diener, E., Lucas, R. E., & Scollon, C. N. (2006). Beyond the hedonic treadmill. *American Psychologist, Vol 61*(No. 4), 305-314.

Dimaggio, G. (2006). Changing the dialogue between self voices during psychotherapy. *Psychotherapy Integration, 16*(No.3), 313-345.

Doidge, N. (2007). *The brain that changes it self: Stories of personal triumph from the frontiers of brain science.* New York: Viking Penguin.

Epstein, S. (1994). Integration of the cognitive and the psychdynamic unconscious. *American Psychologist, 49*(8), 709-724.

Epstein, S. (1998). *Constructive thinking: the key to emotional intelligence.* Westport, Connecticut, London: Praeger.

Epstein, S. (2003). Cognitive-experiential self-theory of personality. In T. Millon & M.J. Lerner (Eds.), *Comprehensive handbook of psychology: volume5: Personality and social psychology* (Vol. 5, pp. 159-184). Hoboken, NJ: Wiley & Sons.

Fonagy, P., Gergely, G., Jurist, E., & Target, M. (2002). *Affect regulation, mentalization and the development of the self.* New York: Other Press.

Fosha, D. (2000). *The transforming power of affect: A model of accelerated change.* New York: Basic Books.

Fosha, D., Siegel, D.J., & Solomon, M. (Eds.). (2009). *The healing power of emotion: Affective neuroscience, development, and clinical practice.* New York: W. W. Norton & Company, Inc.

Fox, D. (1995). Neurodynamics and analytic foundations for Epstein's paradigm. *American Psychologist, 50*(9), 798-799.

Fredrickson, B. L. & Losada, M. (2005). Positive affect and the complex dynamics of human flourishing. *American Psychologist,* 60 (7) 678-686.

Gazzaniga, M. (2008). *Human: The science behind what makes your brain unique.* New York: Harper Perennial.

Gottman, J. (with) Silver, N. (1994) *Why Marriages Succeed or Fail: And How You Can Make Yours Last.* Simon & Schuster.

Grawe, K. (2007). *Neuropsyotherapy: How the neurosciences inform effective psychotherapy.* Mahwah, New Jersey: Lawrence Erlbaum Associates.

Greenberg, M. T., Weissberg, R. P., O'Brian, M. U., Zins, J.. E., Fredericks, L., Resnik, H., et al. (2003). Enhancing school-based prevention and youth development through coordinated social, emotional, and academic learning. *American Psychologist, 58*(6/7), 466-474.

Greenspan, S. I., & Shanker, S. G. (2004). *The first idea: How symbols, language & intelligence evolved from our primitive ancestors to modern humans.* MA: Da Capo Press.

Gross, C. G. (2005). Processing the facial image: A brief history. *American Psychologist, 60*(8), 753-755.

Hebb, D. O. (1949). *The Organization of Behavior: A neuropsychological theory.* New York: Wiley.

Helms, J. E. (2006). Fairness is not validity or cultural bias in racial-group assessment: A quantitative perspective. *American Psychologist, Vol 61*(No.8), 842-845.

Hwang, W. (2006). The psychotherapy adaptation and modification framework: Application to Asian Americans. *American Psychologist, 61*(Number 7), 702-715.

Johnson, S. B., & Millstein, S. J. (2003). Prevention opportunities in health care settings. *American Psychologist, 58*(6/7), 475-481.

Jordan, J.S. (2008, June). Wild agency: Nested intentionalities in cognitive neuroscience and archaeology. *Phil. Trans. R. Soc. B, 363,* 1981-1991.

Jost, J. T. (2006). The end of the end of ideology. *American Psychologist, 61*(Number 7), 651-670.

Kandel, E. R. (2005). *Psychiatry, psyshoanalysis, and the new biology of mind.* American Psychiatric Association Publishing.

Kandel, E. (2006). *In search of memory: The emergence of a new science of mind.* Norton.

Kumpfer, K. L., & Alvarado, R. (2003). Family-strengthening approaches for the prevention of youth problems behaviors. *American Psychologist, 58*(6/7), 457-465.

LeDoux, J. (1996). *The emotional brain: The mysterious underpinning of emotional life.* New York: Simon & Schuster.

LeDoux, J. (2002). *Synaptic self: How our brains become who we are.* England: Penguin Books.

Lewis G. (2009). Are we our brains? *The triple helix Cambridge 800th Anniversary Editions.*

Lynch, Ed; Tucker, Suzi (2005), *Messengers of healing: The family constellations of Bert Hellinger through the eyes of a new generation of practitioners.,* Phoenix: Zeig, Tucker & Theisen.

Main, M., Kaplan, N., & Cassidy, J. (1985). Security in infancy, childhood, and adulthood: A move to the level of representation. *Monographs of the society for research and development, 50,* 66-104.

Main, M., & Solomon, J. (1990). Procedures for identifying infants as disorganized/disoriented during the Ainsworth Strange Situation (pp. 121-160). Chicago: University of Chicago Press.

Masterson, J. (2005). *The personality disorders through the lens of attachment theory and the neurobiologic development of the self.* Arizona: Zeig,Tucker & Theisen, Inc.

Matthews, K. A. (2005). Psychological perspectives on the development of coronary heart disease. *American Psychologist, 60*(8), 780-783.

Melito, R. (2006). Integrating individual and family therapies: A structural developmental approach. *Psychotherapy Integration, Vol.16* (No.3), 346-381.

Michelson, L.K., & Ray, W.J. (Eds.). (1996). *Handbook of dissociation: Theoretical, empirical, and clinical perspectives.* New York: Plenum Press.

Milgram, S. (1974), *Obedience to Authority; An Experimental View.* HarperColins.

Mineka, S., & Zinbarg, R. (2006). A contemporary learning theory perspective on the etiology of anxiety disorders: It's not what you thought it was. *American Psychologist, 61*(1), 10-27.

Moss, R. A. (2007). Negative emotional memories in clinical treatment: Theoretical considerations. *Journal of Psychotherapy Integration, 17*(2), 209-224.

Nation, M., Crusto, C., Wandersman, A., Seybolt, D., Morrissey-Kane, E., & Davino, K. (2003). What works in prevention: Principles of effective prevention programs. *American Psychologist, 58*(6/7), 449-456.

Paivio, A. (1990). *Mental representations: A dual coding approach.* New York: Oxford University Press.

Panksepp, J. (1998). *Affective neuroscience: The foundations of human and animal emotions.* New York: Oxford University Press.

Panksepp, J. (2009). Brain emotional systems and qualities of mental life: From animal models of affect to implications for psychotherapeutics. *The healing power of emotion: Affective neuroscience, development, and clinical practice.* New York: Norton.

Paulsen, S. (1995). EMDR: Its cautious use in the dissociative disorders. *Dissociation, 81*, 32-41.

Perina, K. (2002). Mastering their domain: Who will coach the executive coaches. *Psychology Today,* (December 2002), 15 & 16.

Perls, F. S., Hefferline, R., & Goodman, P. (1951). *Gestalt therapy.* New York: Dell.

Piaget, J. (1952). *The origins of intelligence in children.* New York: International Universities Press.

Porges, S.W. (2009). Reciprocal influences between body and brain in the perception and expression of affects: A polyvagal perspective. *The healing power of emotion: Affective neuroscience, development, and clinical practice.* New York: Norton.

Poultney, R. (2002). Rejection reduces the capacity for intelligent thought. *Psychology Today.*

Renfrew, C., Frith, C., & Malafouris, L. (Eds.). (2008, June). The sapient mind: Archaeology meets neuroscience. Phil. Trans. R. Soc. B, 363, 1935-2061.

Restak, R. (1993) The Brain Has a Mind of Its Own: Insights from a Practicing Neurologist. Random House.

Ripple, C. H., & Zigler, E. (2003). Research, policy, and the federal role in prevention initiatives for children. *American Psychologist, 58*(6/7), 482-490.

Sarno, J. E. (2006). *The divided mind: The epidemic of mindbody disorders.* New York: Harper Collins.

Schloss, J. (2002). Altruism and altruistic love: Science, philosophy & religion in dialogue. *Self Conference.*

Schore, A. (1994). *Affect regulation and the origin of self: The neurobiology of emotional development.* NJ: Lawrence Erlbaum Associates.

Schore, A. N. (2002). Advances in neuropsychoanalysis, attachment theory, and trauma research: Implications for self psychology. *Psychoanalytic Inquiry, 22*(3), 433-484.

Schore, A. (2003). *Affect dysregulation and the disorders of the self.* New York: W W Norton and Co.

Schore, A.N. (2003). *Affect regulation and the repair of the self.* New York: Norton.

Schore, A.N. (2009). *Right brain affect regulation: An essential mechanism of development, trauma, dissociation, and psychotherapy.* New York: Norton.

Schwartz, J. M., & Begley, S. (2002). *The mind and the brain: Neuroplasticity and the power of mental force.* New York: Harper Collins.

Schwartz, R. (1995). *Internal families systems therapy.* New York: Guilford Press.

Seigler, R. S. (2005). Children's learning. *American Psychologist, 60*(8), 767-769.

Seligman, Martin E. P. (1975). *Helplessness: On Depression, Development, and Death.* San Francisco: W.H. Freeman.

Seligman, M. E. P. (2006). Positive psychotherapy. *American Psychologist, Vol. 61* (No. 8), 772-774.

Shapiro, F. (1995). Eye movement desensitization and reprocessing: Basic principles, protocols and procedures. New York: Guilford Press.

Shapiro, F. (2002). EMDR as an integrative psychotherapy approach: Experts of diverse orientations explore the paradigm prism. *APA Press.*

Siegel, D. (1999). *The developing mind: Toward a neurobiology of interpersonal experience.* New York: Guilford Press.

Sigal, J. J. (2004). Studies of unwanted babies. *American Psychologist, 59*(3), 183-184.

Solomon, M., & Siegel, D. (2003). *Healing trauma: Attachment, mind, body and brain.* W.W. Norton & Co.

Springer, S. & Deutsch, G. (2001) *Left Brain, Right Brain, Perspectives From Cognitive Neuroscience (Series of Books in Psychology).* WH Freeman.

Stern, D. N. (1995). *The interpersonal world of the infant: A view from psychoanalysis and developmental psychology.* New York: Basic Books.

Sticker, G., & Gold, J. (2006). *A casebook of integrative psychotherapy.* Washington DC: APA Press.

Tronick, E..Z.. (1998). Dyadically expanded states of consciousness and the process of therapeutic change. *Infant Mental Health J.19*(3), 290-299.

Tronick, E. Z. (2003). "Of course all relationships are Unique": How co-creative processes generate unique mother-infant and patient-therapist relationships and change other relationships. *Psychoanalytic Inquiry,* (23), 473-491.

Van Vugt, Hogan, & Kaiser. Leadership, Followership, and Evolution: Some lessons from the past. American Psychologist, April, 2008.

Vandenbos, GR. (Ed.). (2006, July). *APA dictionary of psychology.* American Psychological Association.

Wallace, B. A., & Shapiro, S. L. (2006). Mental balance and well-being: Building bridges between Buddhism and western psychology. *American Psychologist, 61*(Number 7), 690-701.

Wandersman, A., & Florin, P. (2003). Community interventions and effective prevention. *American Psychologist, 58*(6/7), 441-448.

Watkins, J. G., & Watkins, H. H. (1997). *Ego states: Theory and therapy.* New York: W.W. Norton & Co.

Weissberg, R. P., Kumpfer, K. L., & Seligman, E. P. (2003). Prevention that works for children and youth: An introduction. *American Psychologist, 58*(6/7), 4250433.

Winnicott, D. W. (1965). The maturational process and the facilitation environment. New York: International Universities Press.

Zimbardo, P. (2007) The Lucifer Effect: Understanding How Good People Turn Evil. Random House.

About The Author

Dr Adrianne B. Casadaban

PerformanceAndRealization.com

With more than twenty years of experience, study, and results as a clinical and coaching psychologist, Dr. Casadaban created Performance And Realization in 1996 as a culminating expression of her synthesis of science and practice and a life-long commitment to learning what truly enables individuals to achieve health, happiness, and success.

Dr. Casadaban received a B.A.- Psychology at The Catholic University of America, a Ph. D. in Psychology, Social specialty, at Stony Brook University, a Post Graduate Certificate – Sports Performance Psychology, John F. Kennedy University, and later trained with Creative Dimensions In Management, whose psychological approach became the model for executive coaching. Adrianne is a Licensed Psychologist, Clinical Specialty #PSY7178 in California since January 1982.

Previously, Dr. Casadaban taught Psychology at University of California – Santa Cruz, and taught and conducted research with University of California – San Francisco and Stanford University Medical Center. Dr. Casadaban currently lives in the San Francisco Bay Area with her husband and family.

www.ingramcontent.com/pod-product-compliance
Lightning Source LLC
Chambersburg PA
CBHW052318220526
45472CB00001B/177